Lecture Notes in Mathematics

1472

Torben T. Nielsen

Bose Algebras: The Complex and Real Wave Representations

Springer-Verlag
Berlin Heidelberg New York
London Paris Tokyo
Hong Kong Barcelona
Budapest

Author

Torben T. Nielsen
Mathematical Institute, Århus University
and DIAX Telecommunications A/S
Fælledvej 17, 7600 Struer, Denmark

Mathematics Subject Classification (1980): 81C99, 81D05, 47B47

ISBN 3-540-54041-5 Springer-Verlag Berlin Heidelberg New York
ISBN 0-387-54041-4 Springer-Verlag New York Berlin Heidelberg

Printed in Germany

Printing and binding: Druckhaus Beltz, Hemsbach/Bergstr.
2146/3140-543210 - Printed on acid-free paper

Contents

0.	Introduction	1
1.	The Bose algebra $\Gamma_0\mathcal{H},<,>$	4
2.	Lifting operators to $\Gamma\mathcal{H}$	23
3.	The coherent vectors in $\Gamma\mathcal{H}$	33
4.	The Wick ordering and the Weyl relations	45
5.	Some special operators	53
6.	The complex wave representation	66
7.	The real wave representation	72
8.	Bose algebras of operators	79
9.	Wave representations of $\Gamma(\mathcal{H}+\mathcal{H}^*)$	89
10.	Appendix 1: Halmos' lemma	94
11.	Appendix 2: Gaussian measures	96
12.	References	130
13.	Subject index	132

Introduction

The aim of this paper is to present some results, which are consequences of an algebraization of the concept of Bose–Fock space. Though having their origin and principal area of application in theoretical physics, the usefulness of Bose-Fock spaces extends much further. They appear in digital signal processing [26], filtering theory [8] and in a new formalized description of the process of communication within the human society [22]. Bose–Fock spaces were brought to the attention of mathematicians by Irving E. Segal [15,16,17] and others (cf.[3]) in the sixties.

In this paper both the Bose–Fock space and the accompanying creation and annihilation operator formalism have been combined to form a single well known mathematical object, namely an algebra. Given a Hilbert space $\mathcal{H},\langle,\rangle$, consider the free commutative algebra $\Gamma_0\mathcal{H}$ generated by the linear space $\mathcal{H}$ and a fixed multiplicative unit $\emptyset$, called the vacuum. The scalar product from $\mathcal{H}$ is extended over $\Gamma_0\mathcal{H}$ in such a way that the adjoints to the operators of multiplication by elements in $\mathcal{H}$ are derivations in the algebra. The algebra $\Gamma_0\mathcal{H}$ provided with such an extension of the scalar product is known as the Bose algebra with base $\mathcal{H},\langle,\rangle$ (in physical literature the base space is called "the one–boson–space").

The main advantage of treating the Bose–Fock spaces in this manner is that the mathematics, originated by the study of quantum field theory, becomes available to mathematicians with no background in mathematical physics. In contrast to the situation in quantum physics, which has the Weyl relations as its starting point, the Bose algebra formalism is intuitively clearer for those with no prior knowledge of the subject. It can also be directly compared to an algebra for fermions having as its relations the anti–commutator for the fermion annihilation and creation operators, whereas the Weyl relations and their counterpart within the Clifford algebra, belong to areas of mathematics which are not so closely related. Moreover, a number of important facts difficult to expose in the traditional manner now become easier to understand and work with. In particular the role of complex conjugation can be better understood.

There are also purely mathematical reasons for studying Bose algebras. Introducing free commutative, associative multiplication on a linear space with an inner product, and at the same time extending the inner product in a free manner over the obtained algebra, provide an extra structure without introducing new relations. There are several well known mathematical objects which are actually Bose algebras in a more or less obvious way: 1) the algebra of complex polynomials with scalar product as in [1], 2) the algebra generated by Hermite polynomials (here the operation of multiplication is rather obscure), 3) one can also attach a Bose algebra in a natural manner to the stochastic calculus [24], 4) the original L^2–Fock–space also admits a natural Bose algebra structure.

In [6] the reader will find a way of introducing the Bose–Fock space which is similar to ours, based on the concept of symmetric Hilbert spaces or exponential Hilbert spaces. Here one has, for each $n \in \mathbb{N}$, the so–called n–particle space, which is the space obtained by taking n–fold tensor products of elements of the base space and identifying those tensor products which can be obtained from each other by permuting the order of the elements in the product. The symmetric Hilbert space is constructed from these n–particle spaces, and is in very direct correspondence to the Bose algebra generated by the base space.

One of the best known realizations of the Bose–Fock space is the Segal–Bargmann complex wave representation, cf. [1] and [16], in which the elements are entire functions on the complex plane. In this paper construction of the complex functions involves taking the exponential of elements of the base space cf. [5,20], thus the functions become conjugate–entire. It is worth mentioning that exponentials of the elements of the base space are well known as the so–called coherent vectors, which provide the states for laser beams in quantum optics [5]. The complex wave representation within the appropriate L^2-space has the disadvantage of not being surjective.

In [15] Segal gave a self contained and mathematically complete account of the Schrödinger representation from an abstract point of view; here we shall call it the real wave representation. In this paper the real wave representation is constructed in the following way: a complex conjugation is introduced into the (complex) Hilbert space, and from this a functor is constructed which transforms the complex wave representation into the real one. Contrary to the case of the complex wave representation, the real wave representation constitutes a unitary map.

The functor that transforms the complex wave representation into the real one is an operator whose adjoint corresponds to multiplication by a so–called squeezed state with infinite energy. Squeezed states are a subject of special interest in quantum optics. The first successful experiments with squeezed light are quite recent and the theoretical aspects attract growing attention [13].

It turns out that the so–called normal–product–algebra of

creation and annihilation operators yields a Bose algebra of operators, which we analyse in chapter 8 and 9. The operation of taking the adjoint of an operator can now be taken as the complex conjugation and using this conjugation we construct a real wave representation. This representation produces kernels of operators analogous to the Wigner function [21]. The Bose–Fock space consisting of elements of the representation has a complex conjugation determined by the conjugation of operators described above.

It is worth mentioning that if one reads the book of Louisell [11], it will immediately be clear that the method we apply fits the spirit of application, rigorizing the mathematics without making it too obscure for a theoretical physicist.

Both the complex and the real wave representations have very important probabilistic interpretations, which are provided by gaussian measures on infinite dimensional Hilbert space. These are therefore of principal importance for the theory of Bose algebras. With this in mind and since the reader, who may not already be acquainted with measure theory in infinite dimensional linear spaces, can find it very tiresome to seek the necessary information in the available literature, we have added a short appendix on gaussian measures in Hilbert spaces, based on excellent elementary expositions in [18].

I would like to express my thanks to David Adams, Krista Graversen and Bodil Steengaard for reading parts of the manuscript and for making numerous suggestions and corrections.

Chapter 1: The Bose algebra $\Gamma_0\mathcal{H},\langle,\rangle$

A: The free commutative algebra $\Gamma_0\mathcal{H}$

Let $\mathcal{H},\langle,\rangle$ be a separable Hilbert space with the inner product linear in the second variable. Let then $\Gamma_0\mathcal{H}$ denote the free commutative algebra generated by a multiplicative unit $\emptyset$ (called the vacuum) and the Hilbert space $\mathcal{H}$ (called the base or the one-particle space). We denote by $\mathbb{N}$ the set of positive integers. For $n\in\mathbb{N}$ we define

$$\mathcal{H}_0^n = \text{span}\left\{ a_1a_2\ldots a_n \;\middle|\; a_1,a_2,\ldots,a_n\in\mathcal{H} \right\},$$

where $a_1a_2\ldots a_n$ denotes the free commutative product, fulfilling the additional linearity relation

$$(t\cdot a + b)a_2a_3\ldots a_n = t\cdot a\cdot a_2a_3\ldots a_n + b\cdot a_2a_3\ldots a_n$$

with $a,a_2,a_3,\ldots,a_n,b\in\mathcal{H}$ and $t\in\mathbb{C}$.

We consequently identify elements which, by repeating these linear and commutative operations, can be reduced to the same form. Moreover we set

$$\mathcal{H}_0^0 = \text{span}\,\{\emptyset\} = \mathbb{C}\cdot\emptyset$$

and

$$\Gamma_0\mathcal{H} = \bigoplus_{n=0}^{\infty} \mathcal{H}_0^n = \left\{ \sum_{n=0}^{\infty} f_n \;\middle|\; f_n\in \mathcal{H}_0^n \,,\ \text{where almost all } f_n = 0 \right\}.$$

We make $\Gamma_0\mathcal{H}$ an algebra by defining

the addition : $\displaystyle\sum_{n=0}^{\infty} f_n + \sum_{n=0}^{\infty} g_n = \sum_{n=0}^{\infty} (f_n + g_n)$

the multiplication : $\displaystyle\sum_{n=0}^{\infty} f_n \cdot \sum_{n=0}^{\infty} g_n = \sum_{n=0}^{\infty} \sum_{j+k=n} f_j\cdot g_k$

for every $f_n,g_n \in \mathcal{H}_0^n$ with $n\in\mathbb{N}_0 = \mathbb{N}\cup\{0\}$, and defining

$$\emptyset\cdot f = f\cdot\emptyset = f$$

for $f\in\Gamma_0\mathcal{H}$.

It is an easy exercise to show that addition and multiplication are associative and commutative, thus making $\Gamma_0\mathcal{H}$ a commutative algebra with multiplicative unit $\emptyset$.

We shall use the following notation:

$$\underline{r} = (r_1, r_2, \ldots, r_n) \in \mathbb{N}_0^n \quad ,$$

i.e. $r_k \in \mathbb{N}_0$ for $k=1,2,\ldots,n$, we define

$$|\underline{r}| = r_1 + r_2 + \ldots + r_n$$

$$\underline{r}! = r_1! \cdot r_2! \cdot \ldots \cdot r_n!$$

$$e^{\underline{r}} = e_1^{r_1} \cdot e_2^{r_2} \cdot \ldots \cdot e_n^{r_n} \in \mathcal{H}_0^{|\underline{r}|}$$

$$e^0 = \emptyset \; ,$$

where $\{e_1, e_2, \ldots, e_n\}$ is an orthonormal system in $\mathcal{H}$.

Proposition 1.1A: To every $f \in \mathcal{H}_0^n$ there correspond a $k \in \mathbb{N}$ and an orthonormal system $\{e_1, e_2, \ldots, e_k\}$ in $\mathcal{H}$ such that

$$f \in \operatorname{span}\left\{ e^{\underline{r}} \;\middle|\; \underline{r} \in \mathbb{N}_0^k \text{ with } |\underline{r}| = n \right\} .$$

Proof: It is sufficient to consider $f = a_1 a_2 \ldots a_n$, where $a_1, a_2, \ldots, a_n \in \mathcal{H}$. We define a finite dimensional space $\mathcal{K}$,

$$\mathcal{K} = \operatorname{span}\left\{ a_1, a_2, \ldots, a_n \right\} .$$

Choose an orthonormal basis $\{e_1, e_2, \ldots, e_k\}$ in the space $\mathcal{K}$ with $k = \dim \mathcal{K} \leq n$. Then it is possible to find complex numbers $\{t_j^i\}_{i,j}$ such that

$$a_i = \sum_{j=1}^{k} t_j^i \cdot e_j \quad \text{for } i=1,2,\ldots,n \; ,$$

and we get

$$a_1 \cdot a_2 \cdot \ldots \cdot a_n = \sum_{j_1=1}^{k} \sum_{j_2=1}^{k} \ldots \sum_{j_n=1}^{k} t_{j_1}^1 \cdot t_{j_2}^2 \ldots t_{j_n}^n \cdot e_{j_1} e_{j_2} \ldots e_{j_n}$$

$$= \sum_{|\underline{r}|} s_{\underline{r}} \cdot e^{\underline{r}} \quad \text{with } \underline{r} \in \mathbb{N}_0^k \text{ and } |\underline{r}| = n$$

for some $s_{\underline{r}} \in \mathbb{C}$, which evidently is a sum of the desired type.

In the case of f being a sum of generators for $\mathcal{H}_0^n$, we just choose the space $\mathcal{K}$ sufficiently large.

■

The above argument also verifies the following proposition.

Proposition 1.2A: Let $\mathcal{K}$ denote a finite dimensional space and $\{e_1,e_2,\ldots,e_k\}$ an orthonormal basis in the space $\mathcal{K}$. Then the set

$$\{ e^{\underline{r}} \mid \underline{r} \in \mathbb{N}_0^k \}$$

spans the whole $\Gamma_0\mathcal{K}$.

B: The Bose algebra $\Gamma_0\mathcal{H},<,>$

To be in agreement with the notation employed in physics we will often use $a^+(x)$ for the operator of multiplication by $x\in\mathcal{H}$.

We shall extend the inner product in $\mathcal{H}$ to the whole $\Gamma_0\mathcal{H}$ by demanding for $x \in \mathcal{H}$ the operator $a(x)$, the dual of the operator $a^+(x)$, to be defined on the whole $\Gamma_0\mathcal{H}$, and to be a derivation, i.e. fulfilling the Leibniz rule

$$a(x)(f\cdot g) = g\cdot a(x)f + f\cdot a(x)g \quad \text{for every} \quad f,g\in\Gamma_0\mathcal{H} .$$

We make the inner product uniquely determined by defining

$$<\emptyset,\emptyset> = 1 .$$

The operators $a^+(x)$ and $a(x)$ shall be called the creation and the annihilation operators respectively.

Applying the Leibniz rule, we obtain the recursive formula

$$\begin{aligned}<x_1x_2\cdot\cdot x_m,y_1y_2\cdot\cdot y_n> &= <x_2x_3\cdot\cdot x_m,a(x_1)(y_1y_2\cdot\cdot y_n)> \\ &= <x_2x_3\cdot\cdot x_m,y_2\cdot\cdot y_n a(x_1)y_1 + y_1 a(x_1)(y_2\cdots y_n)> ,\end{aligned}$$

and using the relation

$$a(x_1)(y_k) = <x_1,y_k>\emptyset$$

from lemma 1.1B below we obtain for $n\neq m$

$$<x_1x_2\cdots x_m,y_1y_2\cdots y_n> = 0$$

and for $n=m$

$$\begin{aligned}<x_1x_2\cdot\cdot x_n,y_1y_2\cdots y_n> &= \sum_{k=1}^{n} <x_1,y_k>\cdot<x_2x_3\cdot\cdot x_n,y_1y_2\cdot\cdot y_{k-1}y_{k+1}\cdot\cdot y_n> \\ &= \sum_{\pi} <x_1,y_{\pi_1}>\cdot<x_2,y_{\pi_2}>\cdot\ldots\cdot<x_n,y_{\pi_n}>\end{aligned}$$

where π runs through the set of all permutations of the numbers

$\{1,2,\ldots,n\}$.

Lemma 1.1B: For every $x,y\in\mathcal{H}$ we have

1) $a(x)(\emptyset) = 0$

2) $a(x)(y) = \langle x,y\rangle\emptyset$

Further, for $n,m\in\mathbb{N}$ we have

3) $a(x)(y^n) = n\cdot\langle x,y\rangle\cdot y^{n-1}$

4) $a(x)^m\, y^n = \frac{n!}{(n-m)!}\cdot\langle x,y\rangle^m\cdot y^{n-m}$ for $m\leq n$

Proof: Since

$$\emptyset^2 = \emptyset\cdot\emptyset = \emptyset ,$$

by applying the operator $a(x)$ on both sides we get

$$a(x)(\emptyset^2) = (a(x)\emptyset)\emptyset + \emptyset(a(x)\emptyset) = 2\emptyset\cdot a(x)\emptyset = 2a(x)\emptyset = a(x)\emptyset ,$$

and hence

$$a(x)\emptyset = 0 .$$

For every $n\geq 1$ and $c_1,c_2,\ldots,c_n\in\mathcal{H}$ we get

$$\begin{aligned}\langle a(x)y,c_1c_2\ldots c_n\rangle &= \langle y,x\cdot c_1c_2\ldots c_n\rangle = \langle\emptyset,a(y)(x\cdot c_1c_2\ldots c_n)\rangle\\ &= \langle\emptyset,(a(y)x)c_1c_2\ldots c_n\rangle + \langle\emptyset,x\cdot a(y)(c_1c_2\ldots c_n)\rangle\\ &= \langle\emptyset,(a(y)x)c_1c_2\ldots c_n\rangle + \langle a(x)\emptyset,a(y)(c_1c_2\ldots c_n)\rangle\\ &= \langle a(c_1)\emptyset,c_2c_3\ldots c_n\cdot(a(y)x)\rangle + 0 = 0 .\end{aligned}$$

Thus the element $a(x)y$ is orthogonal to $\mathcal{H}_0^n$ for every $n\in\mathbb{N}$, and we get

$$a(x)y = t\cdot\emptyset \quad\text{for some}\quad t\in\mathbb{C} .$$

As $t = \langle\emptyset,t\cdot\emptyset\rangle = \langle\emptyset,a(x)y\rangle = \langle x,y\rangle$, we have proved identity 2 .

Moreover we have by induction

$$\begin{aligned}a(x)y^n &= a(x)(y\cdot y^{n-1}) = (a(x)y)y^{n-1} + y\cdot(a(x)y^{n-1})\\ &= \langle x,y\rangle y^{n-1} + y\cdot(n-1)\langle x,y\rangle y^{n-2} = n\cdot\langle x,y\rangle y^{n-1} .\end{aligned}$$

The last identity easily follows by induction.

■

We often identify an element and the operator consisting of

multiplication by the element itself, i.e. for $x \in \mathcal{H}$ we shall often use the symbol x for the operator $a^{+}(x)$.

Given a linear operator x , in mathematics one will as a rule write x^{*} for the operator adjoint to the operator x . In this presentation, the annihilation operator is the adjoint to the creation operator. To be able to use intuitions and techniques from operator theory, a mathematician will gladly use this parallel notation when computing. Hence we shall alternatively write x^{*} for the annihilation operator $a(x)$, i.e.

$$a^{+}(x)f = x \cdot f$$

$$a(x)f = x^{*}f$$

for every $f \in \Gamma_{0}\mathcal{H}$.

Proposition 1.2B: For arbitrary $a_1, a_2, \ldots, a_m, b \in \mathcal{H}$ we have the identity

$$\langle a_1 a_2 \ldots a_m, b^n \rangle = \begin{cases} 0 & \text{for } n \neq m \\ n! \cdot \langle a_1, b \rangle \langle a_2, b \rangle \ldots \langle a_n, b \rangle & \text{for } n = m \end{cases} .$$

Proof: The proposition follows by induction and the following calculation.

$$\begin{aligned} \langle a_1 a_2 \ldots a_m, b^n \rangle &= \langle a_2 a_3 \ldots a_m, a(a_1)(b^n) \rangle \\ &= \langle a_2 a_3 \ldots a_m, a_1^{*}(b^n) \rangle = n \cdot \langle a_1, b \rangle \cdot \langle a_2 a_3 \ldots a_m, b^{n-1} \rangle . \end{aligned}$$

■

We are now able to formulate and prove a result, which later on will turn out to be very useful.

Proposition 1.3B: To every $n \in \mathbb{N}$ we have

$$\mathcal{H}_0^n = \operatorname{span} \left\{ a^n \;\middle|\; a \in \mathcal{H} \right\} .$$

Proof: Assuming that $\mathcal{K}$ is a finite dimensional subspace of the Hilbert space $\mathcal{H}$, we will prove that

$$\mathcal{K}_0^n = \text{span}\left\{ a^n \;\middle|\; a\in\mathcal{K} \right\} .$$

Notice that $\mathcal{K}_0^n$ is a finite dimensional subspace as well.

Take $f\in\mathcal{K}_0^n$ and assume that $f \in \left\{ a^n \;\middle|\; a\in\mathcal{K} \right\}^{\perp}$. We have to prove that $f = 0$.

Choose an orthonormal basis $\{e_1,e_2,\ldots,e_k\}$ in the space $\mathcal{K}$ with $k\in\mathbb{N}$. Since the set $\{e^{\underline{r}} \mid \underline{r}\in\mathbb{N}_0^k \text{ with } |\underline{r}| = n \}$ is a basis in $\mathcal{K}_0^n$, we expand in a finite sum

$$f = \sum_{\underline{r}} t_{\underline{r}}\cdot e^{\underline{r}} .$$

For every $a\in\mathcal{K}$ we then have

$$0 = \langle f,a^n\rangle = \sum_{\underline{r}} \overline{t}_{\underline{r}}\cdot\langle e^{\underline{r}},a^n\rangle = \sum_{\underline{r}} \overline{t}_{\underline{r}}\cdot n!\cdot\langle e_1,a\rangle^{r_1}\cdot\langle e_2,a\rangle^{r_2}\ldots\langle e_k,a\rangle^{r_k}$$

$$= \sum_{\underline{r}} \overline{t}_{\underline{r}}\cdot n!\cdot a_1^{r_1}\cdot a_2^{r_2}\ldots a_k^{r_k} ,$$

where $a_i = \langle e_i,a\rangle$ for $i=1,2,\ldots,k$. As a is running through the whole $\mathcal{K}$, the variables $a_1,a_2,\ldots,a_k$ range the whole $\mathbb{C}$, and consequently

$$t_{\underline{r}} = 0 \quad \text{for every } \underline{r} .$$

As every element in $\mathcal{H}_0^n$ is finitely generated, the proposition holds.

∎

The above result should not be surprising, since the general polarization identity

$$n!\cdot x_1x_2\ldots x_n = \sum_{k=1}^{n} (-1)^{n-k}\cdot \sum_{i_1<\ldots<i_k} (x_{i_1}+x_{i_2}+\ldots+x_{i_k})^n$$

for commuting variables $x_1,x_2,\ldots,x_n$ is well known.

Theorem 1.4B: (The canonical commutation relation, cf. [2]) Let as usual $[A,B]$ denote the commutator $AB - BA$ and I the identity operator. For every $x,y\in\mathcal{H}$ the identity

$$[a(x), a^+(y)] = \langle x,y\rangle \cdot I$$

holds on the whole $\Gamma_0\mathcal{H}$.

Proof: This is an easy consequence of the fact that $a(x)$ is a derivation. For $f \in \Gamma_0\mathcal{H}$ we get

$$x^*(y \cdot f) = (x^*y)f + y \cdot (x^*f) = \langle x,y\rangle f + y \cdot (x^*f) .$$

■

Definition 1.5B: A commutative algebra with inner product $\langle,\rangle$, generated by a unit $\emptyset$ (called the vacuum) and a Hilbert space $\mathcal{H},\langle,\rangle$ (called the base) shall be called a Bose algebra if the operator $a(x) = x^*$, dual to the operator $a^+(x)$ of multiplication by $x \in \mathcal{H}$, is defined on the whole algebra, and if the identities

$$a(x)\emptyset = 0$$

$$[a(x), a^+(y)] = \langle x,y\rangle \cdot I$$

are fulfilled for every $x,y \in \mathcal{H}$.

As mentioned earlier the operators $a^+(x)$ are called creation operators, and $a(x)$ are called annihilation operators.

By applying the creation operator $a^+(x)$ to an n–particle state in $\mathcal{H}_0^n$ we get an (n+1)–particle state in $\mathcal{H}_0^{n+1}$ (apart from normalization), and by applying the annihilation operator $a(x)$ to an n–particle state in $\mathcal{H}_0^n$ we get an (n–1)–particle state in $\mathcal{H}_0^{n-1}$ (apart from normalization).

Every Bose algebra then automatically axiomatizes the commutation relation CCR for the creation and annihilation operators of bosons (particles with integer spin). Bosons are particles like photons, phonons, mesons, and the helium nucleus ${}_2^4He$. In particle physics all force exchanging particles are bosons. The photons are the carriers of the electromagnetic force, the π–mesons and the K–mesons are the carriers of the strong interaction and the W^+, W^- and Z^0

particles are the carriers of the weak interaction.

Proposition 1.6B: Consider $x,y\in\mathcal{H}$. The dual $a(x) = x^*$ in the Bose algebra is automatically a derivation.

Proof: We must prove that

$$x^*(f\cdot g) = (x^*f)\cdot g + f\cdot(x^*g) \quad \text{for every } f,g\in\Gamma_0\mathcal{H} .$$

It is sufficient to consider $f = a^n$ for some $a\in\mathcal{H}$ and $n\in\mathbb{N}_0$. The identity

$$x^*(a^n\cdot g) = a^n\cdot x^*g + n\cdot\langle x,a\rangle\cdot a^{n-1}\cdot g$$

is an easy consequence of the commutation relation, proved by induction in the variable $n\in\mathbb{N}$. It remains to show that

$$x^*a^n = n\cdot\langle x,a\rangle\cdot a^{n-1} ,$$

which easily follows by setting $g = \emptyset$ in the above formula and using $x^*\emptyset = 0$.

■

The requirement that for $x,y\in\mathcal{H}$ the operators $a^+(y)$ and $a(x)$ satisfy the relations

$$a(x)\emptyset = 0$$

$$[a(x),a^+(y)] = \langle x,y\rangle I ,$$

can thus – in the definition of a Bose algebra – be replaced by demanding the operator $a(x)$ to be a derivation.

Lemma 1.7B: We shall prove the following assertions, which will be often needed in the sequel.

1) The subspaces $\mathcal{H}_0^n$ are pairwise orthogonal for different indices.
2) Letting $e_1,e_2,\dots,e_n$ be an orthonormal system in $\mathcal{H}$ and defining

$$e^{\underline{r}} = e_1^{r_1} e_2^{r_2} \dots e_n^{r_n} \in \mathcal{H}_0^{|\underline{r}|} \quad \text{for } \underline{r} \in \mathbb{N}_0^n ,$$

we have the inner product

$$\langle e^{\underline{r}},e^{\underline{s}}\rangle = \underline{r}!\cdot\delta_{\underline{r},\underline{s}}$$

for every $\underline{r},\underline{s} \in \mathbb{N}_0^n$.

<u>**Proof:**</u> 1) is a consequence of proposition 1.2B and proposition 1.3B. Regarding 2) we get from the commutation that

$$[e_k^*, a^+(e_i)] = \delta_{i,k} \cdot I \ .$$

By introducing the notation

$$\underline{r} - 1_k = (r_1, r_2, \ldots, r_{k-1}, r_k - 1, r_{k+1}, \ldots, r_n) \ ,$$

we conclude that

$$e_k^*(e^{\underline{r}}) = r_k \cdot e^{(\underline{r}-1_k)} \ .$$

Assuming $\underline{r} = \underline{s}$, we get by induction

$$\langle e^{\underline{r}}, e^{\underline{s}} \rangle = \langle e^{\underline{r}}, e^{\underline{r}} \rangle = \langle e_1^{r_1} e_2^{r_2} \cdots e_n^{r_n}, e^{\underline{r}} \rangle = \langle e^{(\underline{r}-1_1)}, e_1^* e^{\underline{r}} \rangle$$

$$= r_1 \cdot \langle e^{(\underline{r}-1_1)}, e^{(\underline{r}-1_1)} \rangle = r_1! \cdot r_2! \cdot \ldots \cdot r_n! = \underline{r}! \ .$$

If instead we have $\underline{r} \neq \underline{s}$, then there exists a positive integer k fulfilling $r_k \neq s_k$. Without loss of generality we will assume that $r_k < s_k$. Because the operators e_k^* and $a^+(e_i)$ commute for different indices, we get

$$\langle e^{\underline{r}}, e^{\underline{s}} \rangle = \langle e_k^{r_k} \cdot e^{(\underline{r}-r_k)}, e_k^{s_k} \cdot e^{(\underline{s}-s_k)} \rangle$$

$$= \langle a(e_k^{s_k}) e_k^{r_k} \cdot e^{(\underline{r}-r_k)}, e^{(\underline{s}-s_k)} \rangle = 0 \ .$$

■

We write $\Gamma\mathcal{H}$ for the Hilbert space which is the completion of $\Gamma_0\mathcal{H}, \langle , \rangle$.

The following proposition gives an estimate of the growth in norm by multiplication, which is not a bounded operator. We shall use the notation

$$|f| = \langle f,f \rangle^{\frac{1}{2}} \quad \text{for} \quad f \in \Gamma\mathcal{H} \ .$$

<u>**Proposition 1.8B:**</u> To every $f \in \mathcal{H}_0^n$ and $g \in \mathcal{H}_0^m$ we have

$$|f \cdot g| \leq \binom{n+m}{n}^{\frac{1}{2}} \cdot |f| \cdot |g| \ ,$$

with $\binom{n}{m}$ being the binomial coefficient $\frac{n!}{m!(n-m)!}$ for positive integers $n,m\in\mathbb{N}$.

Before stating the proof we shall need a lemma and a concise notation. For $\underline{i},\underline{k}\in\mathbb{N}_0^p$ we define

$$\underline{i}\leq\underline{k} \quad \text{if} \quad i_j\leq k_j \quad \text{for } j=1,2,\ldots,p$$

with $p\in\mathbb{N}$.

Lemma 1.9B: Consider positive integers $m,n\in\mathbb{N}$ fulfilling the inequality $m\leq n$. To every $\underline{k}\in\mathbb{N}_0^p$ with length $|\underline{k}|=n$ we have

$$\sum_{\substack{\underline{i}\leq\underline{k}\\ |\underline{i}|=m}} \frac{\underline{k}!}{\underline{k}!\cdot(\underline{k}-\underline{i})!} = \binom{n}{m} .$$

Notice that the length of the vector $\underline{i}$ does not vary.

Proof: Consider the binomial series of $(s+t)^{k_i}$

$$(s+t)^{k_i} = \sum_{j=0}^{k_i} \binom{k_i}{j}\cdot s^j\cdot t^{k_i-j} .$$

By multiplication of the series for $i=1,2,\ldots,p$ we get

$$(s+t)^n = (s+t)^{k_1+k_2+..+k_p}$$

$$= \sum_{j_1=0}^{k_1}\sum_{j_2=0}^{k_2}\cdots\sum_{j_p=0}^{k_p}\binom{k_1}{j_1}\cdot\binom{k_2}{j_2}\cdots\binom{k_p}{j_p}\cdot s^{j_1+j_2+..+j_p}\cdot$$

$$t^{k_1+k_2+..+k_p-(j_1+j_2+..+j_p)}$$

$$= \sum_{\underline{i}\leq\underline{k}} \frac{\underline{k}!}{\underline{k}!\cdot(\underline{k}-\underline{i})!}\cdot s^{i_1+i_2+..+i_p}\cdot t^{n-(i_1+i_2+..+i_p)} .$$

Notice that the length of the vector $\underline{i}$ varies.

Take the binomial series for $(s+t)^n$,

$$(s+t)^n = \sum_{m=1}^{n}\binom{n}{m}\cdot s^m\cdot t^{n-m} .$$

By collecting the terms involving indices $\underline{i}$ of constant length $|\underline{i}|=m$ in the previous sum, and comparing with the above result, we get

$$\sum_{\substack{\underline{i}\leq\underline{k}\\|\underline{i}|=m}} \frac{\underline{k}!}{\underline{k}!\cdot(\underline{k}-\underline{i})!} = \binom{n}{m} .$$

■

Proof of proposition 1.8B: From proposition 1.2A we find a common orthonormal system $\{e_1,e_2,\dots,e_p\}$ in $\mathcal{H}$ (and matching Fourier coefficients) fulfilling

$$f = \sum_{\underline{r}} a_{\underline{r}}\cdot e^{\underline{r}} \quad \text{for} \quad \underline{r}\in\mathbb{N}_0^p \quad \text{with length} \quad |\underline{r}| = n$$

and

$$g = \sum_{\underline{s}} b_{\underline{s}}\cdot e^{\underline{s}} \quad \text{for} \quad \underline{s}\in\mathbb{N}_0^p \quad \text{with length} \quad |\underline{s}| = m$$

with both sums finite. We then get by multiplication

$$f\cdot g = \sum_{\underline{r}} \sum_{\underline{s}} a_{\underline{r}}\cdot b_{\underline{s}}\cdot e^{\underline{r}+\underline{s}} = \sum_{\underline{k}} \sum_{\underline{i}+\underline{j}=\underline{k}} a_{\underline{i}}\cdot b_{\underline{j}}\cdot e^{\underline{k}}$$

$$= \sum_{\underline{k}} \sum_{\underline{i}\leq\underline{k}} a_{\underline{i}}\cdot b_{\underline{k}-\underline{i}}\cdot e^{\underline{k}}$$

with indices fulfilling

$$|\underline{i}| = n , \quad |\underline{j}| = m \quad \text{and} \quad |\underline{k}| = n+m .$$

Moreover we get

$$|f|^2 = \langle f,f\rangle = \sum_{\underline{r}} \sum_{\underline{r}'} a_{\underline{r}}\cdot \overline{a_{\underline{r}'}}\cdot\langle e^{\underline{r}},e^{\underline{r}'}\rangle = \sum_{\underline{r}} |a_{\underline{r}}|^2\cdot\underline{r}!$$

$$|g|^2 = \sum_{\underline{s}} |b_{\underline{s}}|^2\cdot\underline{s}!$$

and

$$|f\cdot g|^2 = \sum_{\underline{k}} \Big| \sum_{\underline{i}\leq\underline{k}} a_{\underline{i}}\cdot b_{\underline{k}-\underline{i}} \Big|^2\cdot\underline{k}! \quad \text{with} \quad |\underline{i}| = n , \ |\underline{k}| = n+m$$

$$|f|^2\cdot|g|^2 = \sum_{\underline{r}} \sum_{\underline{s}} |a_{\underline{r}}|^2\cdot|b_{\underline{s}}|^2\cdot\underline{r}!\cdot\underline{s}!$$

$$= \sum_{\underline{k}} \sum_{\underline{i}\leq\underline{k}} |a_{\underline{i}}|^2\cdot|b_{\underline{k}-\underline{i}}|^2\cdot\underline{i}!(\underline{k}-\underline{i})! \quad , \quad |\underline{i}| = n, \ |\underline{k}| = n+m .$$

In what follows $|\underline{i}| = n$ and $|\underline{k}| = n+m$. By lemma 1.9B we get

$$\underline{k}!\cdot\left|\sum_{\underline{i}\leq\underline{k}} a_{\underline{i}}\cdot b_{\underline{k}-\underline{i}}\right|^2 = \left|\sum_{\underline{i}\leq\underline{k}} \underline{k}!^{\frac{1}{2}} a_{\underline{i}}\cdot b_{\underline{k}-\underline{i}}\right|^2$$

$$= \left|\sum_{\underline{i}\leq\underline{k}} \left[\frac{\underline{k}!}{\underline{i}!\cdot(\underline{k}-\underline{i})!}\right]^{\frac{1}{2}}\cdot(\underline{i}!\cdot(\underline{k}-\underline{i})!)^{\frac{1}{2}}\cdot a_{\underline{i}}\cdot b_{\underline{k}-\underline{i}}\right|^2$$

$$\leq \sum_{\underline{i}\leq\underline{k}} \frac{\underline{k}!}{\underline{i}!\cdot(\underline{k}-\underline{i})!}\cdot\sum_{\underline{i}\leq\underline{k}} \underline{i}!\cdot(\underline{k}-\underline{i})!\cdot|a_{\underline{i}}|^2\cdot|b_{\underline{k}-\underline{i}}|^2$$

$$= \binom{n}{m}\cdot\sum_{\underline{i}\leq\underline{k}} \underline{i}!\cdot(\underline{k}-\underline{i})!\cdot|a_{\underline{i}}|^2\cdot|b_{\underline{k}-\underline{i}}|^2 .$$

The desired result follows by summation over $\underline{k}$.

■

We notice that the constant $\binom{n+m}{n}$ cannot be improved, as for $a\in\mathcal{H}$ and $m,n\in\mathbb{N}$ we have

$$|a^n\cdot a^m| = |a^{n+m}| = (n+m)!^{\frac{1}{2}}\cdot|a|^{n+m} .$$

Proposition 1.10B: Consider $k\in\mathbb{N}$ and a convergent sequence $\{a_n\}_{n\in\mathbb{N}}$ in $\mathcal{H}$,

$$a_n \xrightarrow[n]{} a \in \mathcal{H} .$$

Then the sequence $\{a_n^k\}_{n\in\mathbb{N}}$ will converge to a^k in $\Gamma\mathcal{H}$,

$$a_n^k \xrightarrow[n]{} a^k \in \Gamma\mathcal{H} .$$

Proof: We start by making an estimate. To every $a,b\in\mathcal{H}$ we have, by using proposition 1.8B several times,

$$|a^k - b^k| = \left|(a-b)\cdot\sum_{j=1}^{k} a^{k-j}\cdot b^{j-1}\right| \leq k^{\frac{1}{2}}\cdot|a-b|\cdot\sum_{j=1}^{k}\left|a^{k-j}\cdot b^{j-1}\right|$$

$$\leq k^{\frac{1}{2}}\cdot|a-b|\cdot\sum_{j=1}^{k}\binom{k-1}{j-1}^{\frac{1}{2}}\cdot|a^{k-j}|\cdot|b^{j-1}|$$

$$\leq k^{\frac{1}{2}}\cdot|a-b|\cdot\sum_{j=1}^{k}\binom{k-1}{j-1}^{\frac{1}{2}}\cdot(k-j)!^{\frac{1}{2}}\cdot|a|^{k-j}\cdot(j-1)!^{\frac{1}{2}}\cdot|b|^{j-1}$$

$$|a^k - b^k| = k^{\frac{1}{2}} \cdot |a - b| \cdot (k-1)!^{\frac{1}{2}} \cdot \sum_{j=1}^{k} |a|^{k-j} \cdot |b|^{j-1}$$

$$\leq k!^{\frac{1}{2}} \cdot |a - b| \cdot \sum_{j=1}^{k} (\max(|a|,|b|))^{k-1}$$

$$\leq k!^{\frac{1}{2}} \cdot |a - b| \cdot k \cdot (|a| + |b|)^{k-1} .$$

Now we easily get the desired result

$$|a^k - a_n^k| \leq k!^{\frac{1}{2}} \cdot k \cdot (|a| + |a_n|)^{k-1} \cdot |a - a_n| \xrightarrow[n\to\infty]{} 0 .$$

■

The above calculated inequality will be frequently used, and therefore we place it as a lemma.

<u>**Lemma 1.11B:**</u> To every $k \in \mathbb{N}$ and $a,b \in \mathcal{H}$ we have

$$|a^k - b^k| \leq k!^{\frac{1}{2}} \cdot k \cdot (|a| + |b|)^{k-1} \cdot |a - b| .$$

<u>**Theorem 1.12B:**</u> (Wiener) Let $\{e_n\}_{n \in \mathbb{N}}$ be an orthonormal basis in $\mathcal{H}$. Consider the set of indices,

$$I = \bigcup_{n=1}^{\infty} \{ \underline{r} \in \mathbb{N}_0^n \} .$$

Then the vectors

$$\left\{ \underline{r}!^{-\frac{1}{2}} \cdot e^{\underline{r}} \right\}_{\underline{r} \in I}$$

form an orthonormal basis in $\Gamma\mathcal{H}$.

<u>**Proof:**</u> It has already been proved that

$$\left\{ \underline{r}!^{-\frac{1}{2}} \cdot e^{\underline{r}} \right\}_{\underline{r} \in I}$$

forms an orthonormal system in $\Gamma\mathcal{H}$. We prove completeness by showing that this orthonormal system is total in $\Gamma_0\mathcal{H}$. Consider $a \in \mathcal{H}$ and $p \in \mathbb{N}$. Then we calculate using proposition 1.10B,

$$a^p = (\lim_n \sum_{k=1}^{n} \langle e_k, a \rangle e_k)^p = \lim_n (\sum_{k=1}^{n} \langle e_k, a \rangle e_k)^p .$$

Since

$$\left(\sum_{k=1}^{n} \langle e_k, a\rangle e_k\right)^p \in \operatorname{span}\left\{ \underline{r}!^{-\frac{1}{2}} \cdot e^{\underline{r}} \right\}_{\underline{r}\in I}$$

and $\left\{ a^p \;\middle|\; a\in\mathcal{H} \text{ and } p\in\mathbb{N}_0 \right\}$ spans the whole $\Gamma_0\mathcal{H}$, the proposition holds.

■

Since the set of indices I is countable, the Hilbert space $\Gamma\mathcal{H}$ is separable.

Definition 1.13B: We define the space $\mathcal{H}^n$ as the closure in $\Gamma\mathcal{H}$ of the space $\mathcal{H}_0^n$.

The spaces $\mathcal{H}^n$ and $\mathcal{H}^m$ are orthogonal subspaces for $n\neq m$.

Lemma 1.14B: Denote by P_n , $n\in\mathbb{N}_0$, the orthogonal projection of $\Gamma\mathcal{H}$ onto $\mathcal{H}^n$. Defining

$$f_{|n} = P_n(f)$$

for $f\in\Gamma\mathcal{H}$, we have

$$f = \sum_{n=0}^{\infty} f_{|n} \quad \text{in } \Gamma\mathcal{H} .$$

Proof: To arbitrary $\epsilon>0$ choose $g\in\Gamma_0\mathcal{H}$ fulfilling

$$|f - g| < \epsilon/2 .$$

As $g\in\Gamma_0\mathcal{H}$, it is possible to find $N\in\mathbb{N}$ and $g_n\in\mathcal{H}_0^n$ such that

$$g = \sum_{n=0}^{N} g_n .$$

Defining $g_n = 0$ for $n>N$, we have for $n>N$ that

$$\left|f - \sum_{k=0}^{n} f_{|k}\right| \le |f - g| + \left|g - \sum_{k=0}^{n} f_{|k}\right|$$

$$\left| f - \sum_{k=0}^{n} f_{|k} \right| < \epsilon/2 + \left| \sum_{k=0}^{n} (g_k - f_{|k}) \right| ,$$

and because $P_k g = P_k(\sum_{n=0}^{N} g_n) = g_k$,

$$\left| f - \sum_{k=0}^{n} f_{|k} \right| \leq \epsilon/2 + \left| \sum_{k=0}^{n} (P_k g - P_k f) \right|$$

$$= \epsilon/2 + \left| (\sum_{k=0}^{n} P_k)(g - f) \right|$$

$$\leq \epsilon/2 + |g - f| < \epsilon/2 + \epsilon/2 = \epsilon .$$

■

As a consequence, we have expressed $\Gamma\mathcal{H}$ as an infinite direct sum of orthogonal closed subspaces $\mathcal{H}^n$.

__Corollary 1.15B:__ $\Gamma\mathcal{H} = \left\{ \sum_{n=0}^{\infty} f_n \;\middle|\; f_n \in \mathcal{H}^n \text{ with } \sum_{n=0}^{\infty} |f_n|^2 < \infty \right\}$.

__Proof:__ This is an easy consequence of the completeness of the space $\Gamma\mathcal{H}$ and lemma 1.14B.

■

From proposition 1.8B we know that the multiplication is __not__ a continuous operation in $\Gamma_0\mathcal{H}$ and therefore cannot be extended to the whole $\Gamma\mathcal{H}$, $\Gamma\mathcal{H}$ thus becoming an algebra. Still it is possible to extend the multiplication to a broader class of elements than those of $\Gamma_0\mathcal{H}$. For fixed numbers $n,m \in \mathbb{N}_0$ the mapping

$$\mathcal{H}_0^n \times \mathcal{H}_0^m \ni (f,g) \longrightarrow f\cdot g \in \mathcal{H}_0^{n+m}$$

is continuous according to proposition 1.8B, and thus it can be extended uniquely to a commutative multiplication

$$\mathcal{H}^n \times \mathcal{H}^m \ni (f,g) \longrightarrow f\cdot g \in \mathcal{H}^{n+m} .$$

Definition 1.16B: Consider $f,g\in\Gamma\mathcal{H}$. According to lemma 1.14B it is possible to write f,g in the form

$$f = \sum_{n=0}^{\infty} f_n \quad \text{with} \quad f_n\in\mathcal{H}^n$$

$$g = \sum_{n=0}^{\infty} g_n \quad \text{with} \quad g_n\in\mathcal{H}^n .$$

We define the product $f\cdot g$ by the series

$$f\cdot g = \sum_{n=0}^{\infty} \sum_{j+k=n} f_j\cdot g_k \in \Gamma\mathcal{H}$$

provided

$$\sum_{n=0}^{\infty} \Big|\sum_{j+k=n} f_j\cdot g_k\Big|^2 < \infty .$$

This new product is clearly an extension of the multiplication in $\Gamma_0\mathcal{H}$, and it is still both commutative and associative when defined.

Lemma 1.17B: Let $\{e_n\}_{n\in\mathbb{N}}$ denote an orthonormal basis in the Hilbert space $\mathcal{H}$. By expanding $f,g\in\Gamma_0\mathcal{H}$ in the base from theorem 1.12B, i.e.

$$f = \sum_{\underline{r}} a_{\underline{r}}\cdot e^{\underline{r}} \in \Gamma_0\mathcal{H}$$

$$g = \sum_{\underline{s}} b_{\underline{s}}\cdot e^{\underline{s}} \in \Gamma_0\mathcal{H} ,$$

we get

$$f\cdot g = \sum_{\underline{n}} \Big(\sum_{\underline{j}+\underline{k}=\underline{n}} a_{\underline{j}}\cdot b_{\underline{k}}\Big)\cdot e^{\underline{n}} \in \Gamma_0\mathcal{H} .$$

Proof: Notice that the indices run through countable sets. We have that

$$\underline{r}!\cdot a_{\underline{r}} = \langle e^{\underline{r}}, f\rangle$$

$$\underline{s}!\cdot b_{\underline{s}} = \langle e^{\underline{s}}, g\rangle\ .$$

$$\langle e^{\underline{n}}, f\cdot g\rangle = \langle f^{*}e^{\underline{n}}, g\rangle = \sum_{\underline{s}} \langle f^{*}e^{\underline{n}}, b_{\underline{s}}e^{\underline{s}}\rangle = \sum_{\underline{s}} b_{\underline{s}}\cdot\langle e^{\underline{n}}, e^{\underline{s}}\cdot f\rangle$$

$$= \sum_{\underline{s}}\sum_{\underline{r}} b_{\underline{s}}\cdot\langle (e^{\underline{s}})^{*}e^{\underline{n}}, a_{\underline{r}}\cdot e^{\underline{r}}\rangle = \sum_{\underline{s}}\sum_{\underline{r}} a_{\underline{r}}\cdot b_{\underline{s}}\cdot\langle e^{\underline{n}}, e^{\underline{r}+\underline{s}}\rangle$$

$$= \sum_{\underline{r}+\underline{s}=\underline{n}} a_{\underline{r}}\cdot b_{\underline{s}}\cdot \underline{n}!\ .$$

Since $f\cdot g\in\Gamma_0\mathcal{H}$, by the theorem of Wiener (theorem 1.12B) we get

$$f\cdot g = \sum_{\underline{n}} c_{\underline{n}}\cdot e^{\underline{n}}\ ,$$

where we have shown that

$$c_{\underline{n}} = \sum_{\underline{r}+\underline{s}=\underline{n}} a_{\underline{r}}\cdot b_{\underline{s}}\ .$$

■

Example 1.18B: Let $n\in\mathbb{N}$ be fixed. For $a_1, a_2, \ldots, a_n\in\mathbb{C}$, consider the conjugate linear function

$$\underline{a} : \mathbb{C}^n \longrightarrow \mathbb{C}$$

defined by its values on vectors of the form $\underline{z} = (z_1, z_2, \ldots, z_n)$ as

$$\underline{a}[\underline{z}] = a_1\bar{z}_1 + a_2\bar{z}_2 + \ldots + a_n\bar{z}_n\ .$$

We define the Hilbert space

$$\mathcal{H} = \left\{ \underline{a}[\cdot] \;\middle|\; \underline{a} = (a_1, a_2, \ldots, a_n)\in\mathbb{C}^n \right\}$$

with the inner product

$$\langle\, \underline{a}[\cdot], \underline{b}[\cdot]\,\rangle = \sum_{k=1}^{n} \bar{a}_k\cdot b_k\ .$$

By treating the integration over $\mathbb{C}$ as integration over $\mathbb{R}^2$ we can easily prove the following statement:

The Bose algebra $\Gamma_0\mathcal{H}$ consists of polynomials in the variables $\bar{z}_1, \bar{z}_2, \ldots, \bar{z}_n\in\mathbb{C}$, and the vacuum $\varnothing$ as the constant function 1 . We have the inner product

$$< \underline{a}[\cdot],\underline{b}[\cdot] > = \pi^{-n}\cdot\int_{\mathbb{C}^n} \overline{\underline{a}[\underline{z}]}\cdot\underline{b}[\underline{z}]\cdot\exp(-|\underline{z}|^2)d\underline{z} \ ,$$

where
$$\underline{z} = (z_1,z_2,\ldots,z_n)\in\mathbb{C}^n$$
and
$$z_k = x_k+i\cdot y_k \quad \text{for} \quad k=1,2,..,n$$
$$|\underline{z}|^2 = |z_1|^2 + |z_2|^2 +\ldots+ |z_n|^2$$
$$d\underline{z} = dx_1dy_1dx_2dy_2\ldots dx_ndy_n \ .$$

By writing z_k, $k\in\mathbb{N}$, for the function
$$\underline{z} \longrightarrow z_k \ ,$$
we then observe that the creation operator $a^+(z_k)$ consists of multiplication by the variable z_k, and that the annihilation operator $a(z_k)$ consists of differentiation with respect to the variable $\bar{z}_k$, i.e.
$$a(z_k) = \frac{\partial}{\partial\bar{z}_k} \ .$$

Example 1.19B: Consider the Hilbert space $\mathcal{H} = \mathbb{C}^n$, $n\in\mathbb{N}$, equipped with the usual scalar product. Define
$$\varnothing(\underline{x}) = \pi^{-n/4}\cdot\exp(-\tfrac{1}{2}\|\underline{x}\|^2) : \mathbb{R}^n \longrightarrow \mathbb{C} \ ,$$
and denote by $\Gamma_0\mathbb{C}^n$ the space of all functions of the form $g(\cdot)\varnothing(\cdot)$, where $g(\cdot)$ is a complex-valued polynomial on $\mathbb{R}^n$.

To every $\underline{c} = (c_1,c_2,\ldots,c_n) \in \mathbb{C}^n$ assign the operators
$$a(\underline{c}) : \Gamma_0\mathbb{C}^n \longrightarrow \Gamma_0\mathbb{C}^n$$
and
$$a^+(\underline{c}) : \Gamma_0\mathbb{C}^n \longrightarrow \Gamma_0\mathbb{C}^n$$
setting for $f \in \Gamma_0\mathbb{C}^n$
$$a(\underline{c})f(\underline{x}) = 2^{-\frac{1}{2}}\cdot\sum_{k=1}^{n} \bar{c}_k(x_k + \frac{\partial}{\partial x_k})f(\underline{x})$$
and
$$a^+(\underline{c})f(\underline{x}) = 2^{-\frac{1}{2}}\cdot\sum_{k=1}^{n} c_k(x_k - \frac{\partial}{\partial x_k})f(\underline{x}) \ .$$

We easily calculate for $\underline{b},\underline{c}\in\mathbb{C}^n$ that

$$a(\underline{b})\,\emptyset = 0$$

$$\langle a^+(\underline{b})\emptyset, a^+(\underline{c})\emptyset\rangle = \langle \underline{b},\underline{c}\rangle \ ,$$

where the inner product on the left side is the standard inner product in the Hilbert space $L^2(\mathbb{R}^n)$ and on the right side is the standard inner product in $\mathbb{C}^n$, both inner products being conjugate linear in the first variable.

In this way we have isometrically embedded $\mathbb{C}^n$ into $L^2(\mathbb{R}^n)$ by the mapping

$$\mathbb{C}^n \ni \underline{c} \longrightarrow a^+(\underline{c})\emptyset \in L^2(\mathbb{R}^n) \ .$$

By defining the Wick product between $\underline{c}_1, \underline{c}_2, \ldots, \underline{c}_m \in \mathbb{C}^n$ as

$$\underline{c}_1 \cdot \underline{c}_2 \cdot \ldots \cdot \underline{c}_m = a^+(\underline{c}_1) a^+(\underline{c}_2) \ldots a^+(\underline{c}_m)\,\emptyset \ ,$$

we obtain an isometric realization of the Bose algebra $\Gamma_0\mathbb{C}^n$ within the Hilbert space $L^2(\mathbb{R}^n)$. It is well known that the completion of $\Gamma_0\mathbb{C}^n$ which amounts to its closure in $L^2(\mathbb{R}^n)$ is equal to the whole $L^2(\mathbb{R}^n)$. In particular, given natural numbers $k_1, k_2, \ldots, k_n$, we get the Hermite functions as

$$e_1^{k_1} \cdot e_2^{k_2} \cdot \ldots \cdot e_n^{k_n} \ ,$$

where $e_i = (0,\ldots,0,1,0,\ldots,0) \in \mathbb{C}^n$ and the digit 1 is situated at position number i.

Chapter 2: Lifting operators to $\Gamma\mathcal{H}$

In this chapter we shall examine possibilities of extending linear operators between base spaces to homomorphisms of the corresponding Bose algebras.

Proposition 2.1: Let $\mathcal{H}_1$ and $\mathcal{H}_2$ denote two Hilbert spaces, and let

$$A : \mathcal{H}_1 \longrightarrow \mathcal{H}_2$$

be a linear mapping, which is an isometry. Then there exists a uniquely determined linear isometry

$$\Gamma A : \Gamma\mathcal{H}_1 \longrightarrow \Gamma\mathcal{H}_2$$

fulfilling the properties

1) ΓA is an extension of A .
2) The restriction

$$\Gamma A|_{\Gamma_0\mathcal{H}_1} : \Gamma_0\mathcal{H}_1 \longrightarrow \Gamma_0\mathcal{H}_2$$

is a homomorphism of the concerned algebras.

Proof: Let $\{e_n\}_{n\in\mathbb{N}}$ denote an orthonormal basis in the space $\mathcal{H}_1$. Then we observe that $\{Ae_n\}_{n\in\mathbb{N}}$ is an orthonormal system in $\mathcal{H}_2$.

First we will prove that if ΓA is a linear isometry and 2) holds, then ΓA is automatically uniquely determined. Indeed, ΓA being a homomorphism on $\Gamma_0\mathcal{H}_1$ yields that for $x\in\mathcal{H}_1$ and $n\in\mathbb{N}$

$$\Gamma A(x^n) = (\Gamma Ax)^n = (Ax)^n .$$

Since powers of elements from $\mathcal{H}_1$ generate the whole algebra $\Gamma_0\mathcal{H}_1$, ΓA is uniquely determined on the whole $\Gamma_0\mathcal{H}_1$. Since ΓA is also an isometry, ΓA is, by continuity, uniquely determined on $\Gamma\mathcal{H}_1$. Notice that according to the definition of a homomorphism, we demand that ΓA is unit preserving, i.e.

$$\Gamma A(\emptyset_1) = \emptyset_2 ,$$

where $\emptyset_1$ and $\emptyset_2$ are the vacuums in the algebras $\Gamma_0\mathcal{H}_1$ and $\Gamma_0\mathcal{H}_2$

respectively.

We shall show the existence of the operator ΓA . First we define ΓA on the basis $\{e^{\underline{r}}\}_{\underline{r}}$ from theorem 1.12B, setting

$$\Gamma A(e^{\underline{r}}) = (Ae_1)^{r_1}\cdot(Ae_2)^{r_2}\cdot\ldots\cdot(Ae_n)^{r_n} ,$$

then we extend ΓA by linearity and continuity to an isometry on the whole $\Gamma\mathcal{H}_1$.

We have to show that the elements $\Gamma A(f)$ for $f\in\Gamma_0\mathcal{H}_1$ belong to the algebra $\Gamma_0\mathcal{H}_2$. To do so, it is sufficient to consider elements $f = a^p$ for $p\in\mathbb{N}$ and $a\in\mathcal{H}_1$. According to proposition 1.10B we get

$$a^p = \lim_{k\to\infty} a_k^p$$

with $a_k = \sum_{i=1}^{k} \langle e_i,a\rangle e_i$. Since ΓA is an isometry, we have

$$\Gamma A(a^p) = \Gamma A(\lim_{k\to\infty} a_k^p) = \lim_{k\to\infty} \Gamma A(a_k^p) = \lim_{k\to\infty} \Gamma A\left(\left[\sum_{i=1}^{k} \langle e_i,a\rangle e_i\right]^p\right)$$

$$= \lim_{k\to\infty} \left(\left[\sum_{i=1}^{k} \langle e_i,a\rangle Ae_i\right]^p\right) = \lim_{k\to\infty} \left(\left[Aa_k\right]^p\right)$$

$$= (\lim_{k\to\infty} Aa_k)^p = (Aa)^p \in \Gamma_0\mathcal{H}_2 .$$

Hence we have shown the existence of a linear isometry ΓA , extending A , such that

$$\Gamma A(\Gamma_0\mathcal{H}_1) \subset \Gamma_0\mathcal{H}_2 .$$

It remains to show that the operator ΓA is multiplicative. Consider $f,g\in\Gamma_0\mathcal{H}_1$ and take their expansions,

$$f = \sum_{\underline{r}} a_{\underline{r}}\cdot e^{\underline{r}}$$

$$g = \sum_{\underline{s}} b_{\underline{s}}\cdot e^{\underline{s}} ,$$

according to theorem 1.12B. According to lemma 1.17B we have

$$f\cdot g = \sum_{\underline{n}} \sum_{\underline{j}+\underline{k}=\underline{n}} a_{\underline{j}}\cdot b_{\underline{k}}\cdot e^{\underline{n}} .$$

$$\Gamma A(f\cdot g) = \sum_{\underline{n}} \sum_{\underline{j}+\underline{k}=\underline{n}} a_{\underline{j}}\cdot b_{\underline{k}}\cdot \Gamma A(e^{\underline{n}}) = \sum_{\underline{n}} \sum_{\underline{j}+\underline{k}=\underline{n}} a_{\underline{j}}\cdot b_{\underline{k}}\cdot (Ae_1)^{n_1}\ldots(Ae_i)^{n_i} ,$$

and again by lemma 1.17B

$$\Gamma A(f)\cdot\Gamma A(g) = \sum_{\underline{r}} a_{\underline{r}}\cdot(Ae_1)^{r_1}\dots(Ae_i)^{r_i} \cdot \sum_{\underline{s}} b_{\underline{s}}\cdot(Ae_1)^{s_1}\dots(Ae_i)^{s_i}$$

$$= \sum_{\underline{n}} \sum_{\underline{j}+\underline{k}=\underline{n}} a_{\underline{j}}\cdot b_{\underline{k}}\cdot(Ae_1)^{n_1}\dots(Ae_i)^{n_i} .$$

Hence ΓA is multiplicative on $\Gamma_0\mathcal{H}_1$, and the proposition follows.

∎

It is not hard to see that the proposition holds for conjugate linear isometries as well.

Denoting by i the operator of multiplication by the imaginary unit i , the operator Γi is called the (inverse) Fourier transformation. Indeed, if $\Gamma\mathcal{H} = L^2(\mathbb{R}^n)$ and if the operators of creation and annihilation are the usual ones, Γi coincides with the traditional inverse Fourier transformation. Then the spaces $\mathcal{H}^n$ for $n\in\mathbb{N}_0$ are the eigenspaces for the Fourier transformation.

<u>**Proposition 2.2:**</u> Consider a closed subspace $\mathcal{K}$ of $\mathcal{H}$. Let P denote the orthogonal projection of $\mathcal{H}$ onto $\mathcal{K}$ and ΓP the orthogonal projection of $\Gamma\mathcal{H}$ onto $\Gamma\mathcal{K}$. The restriction

$$\Gamma P|_{\Gamma_0\mathcal{H}}$$

is a homomorphism of the concerned algebras extending P .

<u>**Proof:**</u> It is sufficient to show that

$$\Gamma P(x_1x_2\dots x_n) = (Px_1)(Px_2)\dots(Px_n)$$

for $n\in\mathbb{N}$ and $x_1,x_2,\dots,x_n\in\mathcal{H}$. Writing

$$x_i = y_i + z_i \quad \text{for } i=1,2,\dots,n$$

with $y_i\in\mathcal{K}$ and $z_i\in\mathcal{K}^\perp$, the product assumes the form

$$x_1x_2\dots x_n = (y_1 + z_1)\cdot(y_2 + z_2)\cdot\dots\cdot(y_n + z_n)$$
$$= y_1y_2\dots y_n + \sum_i z_i\cdot f_i ,$$

where $f_i \in \Gamma_0 \mathcal{H}$. We evaluate

$$\Gamma P(x_1 x_2 \cdots x_n) = \Gamma P(y_1 y_2 \cdots y_n + \sum_i z_i \cdot f_i)$$
$$= \Gamma P(y_1 y_2 \cdots y_n) + \sum_i \Gamma P(z_i \cdot f_i) \ .$$

Since $z_i \in K^\perp$, for $a \in K$ and every $m \in \mathbb{N}_0$ we get

$$<z_i f_i, a^m> = <f_i, z_i^*(a^m)> = m \cdot <z_i, a> \cdot <f_i, a^{m-1}> = 0 \ ,$$

or, in other words, $z_i \cdot f_i \in (\Gamma K)^\perp$ resulting in $\Gamma P(z_i \cdot f_i) = 0$. Since $y_1 y_2 \cdots y_n \in \Gamma K$ we get

$$\Gamma P(x_1 x_2 \cdots x_n) = \Gamma P(y_1 y_2 \cdots y_n) = y_1 y_2 \cdots y_n$$
$$= (Px_1)(Px_2) \cdots (Px_n) \ .$$

■

Definition 2.3: Let $A : \mathcal{H}_1 \longrightarrow \mathcal{H}_2$ denote a linear transformation between two Hilbert spaces. If the operator A can be uniquely extended to a homomorphism $\Gamma_0 \mathcal{H}_1 \longrightarrow \Gamma_0 \mathcal{H}_2$ we will denote the extension by the symbol ΓA . If additionally ΓA is continuous, its extension over the whole $\Gamma \mathcal{H}_1$ will be denoted by the same symbol.

Notice that the uniqueness of the extension follows automatically from the properties of a homomorphism.

We will prove that if $A \in \mathcal{B}(\mathcal{H}_1, \mathcal{H}_2)$ = the space of all bounded linear transformations between $\mathcal{H}_1$ and $\mathcal{H}_2$, then the operator ΓA exists, and if additionally A is a contraction, i.e. $\|A\| \leq 1$, then ΓA becomes a contraction, defined on the whole $\Gamma \mathcal{H}_1$.

Lemma 2.4: Denote by A_1 and A_2 two linear transformations fulfilling

$$\mathcal{H}_1 \xrightarrow{A_1} \mathcal{H}_2 \xrightarrow{A_2} \mathcal{H}_3 \ .$$

If both ΓA_1 and ΓA_2 exist, then $\Gamma(A_2 A_1)$ exists as well and

$$\Gamma(A_2 A_1) = (\Gamma A_2)(\Gamma A_1) \ .$$

Proof: Since ΓA_1 and ΓA_2 both are homomorphisms, $(\Gamma A_2)(\Gamma A_1)$ is a homomorphism as well. Thus the existence of $\Gamma(A_2A_1)$ is proved.

■

From Halmos' lemma (cf. appendix 1) we directly obtain the following corollary.

Corollary 2.5: Any contraction A can be represented as the composition of a projection P and an isometry Z, i.e

$$A = P \cdot Z .$$

Theorem 2.6: (Nelson, cf. [14]) Let $A : \mathcal{H}_1 \longrightarrow \mathcal{H}_2$ denote a linear contraction. Then it is possible to extend A to a uniquely determined linear contraction

$$\Gamma A : \Gamma\mathcal{H}_1 \longrightarrow \Gamma\mathcal{H}_2$$

such that it maps $\Gamma_0\mathcal{H}_1$ into $\Gamma_0\mathcal{H}_2$, and the restriction

$$\Gamma A|_{\Gamma_0\mathcal{H}_1}$$

is a homomorphism of the respective algebras.

Proof: By corollary 2.5 we write A in the form

$$A = P \cdot Z$$

with

$$P : \mathcal{H}_2 \oplus \mathcal{H}_1 \longrightarrow \mathcal{H}_2$$

being the projection of $\mathcal{H}_2 \oplus \mathcal{H}_1$ onto $\mathcal{H}_2 = \mathcal{H}_2 \oplus 0$, and

$$Z : \mathcal{H}_1 \longrightarrow \mathcal{H}_2 \oplus \mathcal{H}_1$$

being an isometry.

We then extend the map Z to an isometry

$$\Gamma Z : \Gamma\mathcal{H}_1 \longrightarrow \Gamma(\mathcal{H}_2 \oplus \mathcal{H}_1) ,$$

the restriction $\Gamma Z|_{\Gamma_0\mathcal{H}_1}$ of which is a homomorphism. Moreover, let

$$\Gamma P : \Gamma(\mathcal{H}_2 \oplus \mathcal{H}_1) \longrightarrow \Gamma\mathcal{H}_2$$

denote the orthogonal projection of $\Gamma(\mathcal{H}_2 \oplus \mathcal{H}_1)$ onto $\Gamma\mathcal{H}_2$, where the restriction $\Gamma P|_{\Gamma_0(\mathcal{H}_2 \oplus \mathcal{H}_1)}$ is a homomorphism.

Clearly the operator $(\Gamma P)(\Gamma Z)$ is a contraction defined on the whole $\Gamma\mathcal{H}_1$, extending A, and

$$(\Gamma P)(\Gamma Z)|_{\Gamma_0\mathcal{H}_1}$$

is a homomorphism as well.

■

<u>**Corollary 2.7:**</u> Let $A : \mathcal{H}_1 \longrightarrow \mathcal{H}_2$ denote a linear bounded transformation. Then it is possible to extend A uniquely to a homomorphism

$$\Gamma A : \Gamma_0\mathcal{H}_1 \longrightarrow \Gamma_0\mathcal{H}_2 .$$

<u>**Proof:**</u> Uniqueness is a consequence of ΓA being a homomorphism. The operator $A/\|A\|$ is a contraction, which we easily extend to a homomorphism defined on $\Gamma_0\mathcal{H}_1$. It remains to extend the operator consisting of multiplication by $\|A\|$ defined on $\mathcal{H}_2$.

For arbitrary $\lambda \in \mathbb{C}$ consider the operator

$$M_\lambda : \mathcal{H}_2 \longrightarrow \mathcal{H}_2$$

consisting of multiplication by $\lambda \in \mathbb{C}$. We define the extension

$$\Gamma\lambda : \Gamma_0\mathcal{H}_2 \longrightarrow \Gamma_0\mathcal{H}_2$$

on the elements $f = \sum_{n=0}^{\infty} f_{|n} \in \Gamma_0\mathcal{H}_2$ by setting

$$\Gamma\lambda(f) = \sum_{n=0}^{\infty} \lambda^n \cdot f_{|n} ,$$

i.e. the same as defining

$$\Gamma\lambda(x_1 x_2 \ldots x_n) = \lambda^n \cdot x_1 x_2 \ldots x_n$$

for $n \in \mathbb{N}$ and $x_1, x_2, \ldots, x_n \in \mathcal{H}_2$. Then $\Gamma\lambda$ is a well defined homomorphism of $\Gamma_0\mathcal{H}_2$ into itself.

The operator $\Gamma(\|A\|)\cdot\Gamma(A/\|A\|)$ constitutes, of course, a homomorphism extending A .

■

We notice that in the case $\|A\|>1$ the operator ΓA is an unbounded, densely defined operator in $\Gamma\mathcal{H}_1$ with values in $\Gamma\mathcal{H}_2$.

We have now accomplished the task formulated at the beginning of the chapter, but before leaving the subject, we will use the results in a special case.

Consider a strongly continuous one-parameter group of unitary operators

$$\{U_t\}_{t\in\mathbb{R}}$$

with $U_t : \mathcal{H} \longrightarrow \mathcal{H}$. According to Stone's theorem, there exists a densely defined self-adjoint generator A fulfilling

$$U_t = e^{i\cdot t\cdot A} .$$

Using proposition 2.1 we extend every U_t , $t\in\mathbb{R}$, to an isometry, defined on the whole $\Gamma\mathcal{H}$.

<u>**Proposition 2.8:**</u> Define

$$\mathcal{U}_t = \Gamma U_t \text{ for every } t\in\mathbb{R} .$$

Then $\{\mathcal{U}_t\}_{t\in\mathbb{R}}$ is a strongly continuous one-parameter group of unitary operators, defined on $\Gamma\mathcal{H}$.

<u>**Proof:**</u> To every $s,t\in\mathbb{R}$ we have

$$(\Gamma U_s)(\Gamma U_t) = \Gamma(U_{s+t}) ,$$

thereby getting

$$\mathcal{U}_s\mathcal{U}_t = \mathcal{U}_{s+t} ,$$

i.e. $\{\mathcal{U}_t\}_{t\in\mathbb{R}}$ is a group.

As $\mathcal{U}_t$ is an isometry, $\mathcal{U}_t^*\mathcal{U}_t = I$. The existence of $\mathcal{U}_t^{-1}$ ascertains that $\mathcal{U}_t$ is unitary.

Because $\|\mathcal{U}_t\| = 1$, it is sufficient to show that $\mathcal{U}_t$ is strongly continuous at $t=0$. It is also sufficient to show the continuity on a dense subset of $\Gamma\mathcal{H}$. Take $x\in\mathcal{H}$ and $n\in\mathbb{N}$. By the estimate of lemma 1.11B we get

$$|\mathcal{U}_t x^n - x^n| = |(U_t x)^n - x^n| \leq (n!)^{\frac{1}{2}} \cdot n \cdot (|U_t x| + |x|)^{n-1} \cdot |U_t x - x|$$
$$= (n!)^{\frac{1}{2}} \cdot n \cdot (2 \cdot |x|)^{n-1} \cdot |U_t x - x| .$$

By strong continuity of U_t the proposition holds.

■

Definition 2.9: By Stone's theorem there exists a densely defined self-adjoint generator $d\Gamma A$ fulfilling

$$\mathcal{U}_t = e^{i \cdot t \cdot d\Gamma A} .$$

We will call $d\Gamma A$ **the second quantization** of A.

On $\Gamma\mathcal{H}$ we have the operator identity

$$\Gamma(e^{i \cdot t \cdot A}) = e^{i \cdot t \cdot d\Gamma A} .$$

As the operator $\Gamma(e^{i \cdot t \cdot A})$ extends $e^{i \cdot t \cdot A}$, the domain for the operator A, called $\mathcal{D}(A)$, has to be a subset of $\mathcal{D}(d\Gamma A)$,

$$\mathcal{D}(A) \subset \mathcal{D}(d\Gamma A)$$

and then the operator $d\Gamma A$ is an extension of A, i.e.

$$d\Gamma A(x) = A(x) \quad \text{for every} \quad x\in\mathcal{D}(A) .$$

Proposition 2.10: Considered in $\Gamma_0\mathcal{H} \cap \mathcal{D}(d\Gamma A)$, the operator $d\Gamma A$ is a derivation, i.e. it fulfills the Leibniz rule.

Proof: Consider the elements $f,g \in \Gamma_0\mathcal{H} \cap \mathcal{D}(d\Gamma A)$ with the property that the product $f \cdot g \in \mathcal{D}(d\Gamma A)$. Since $e^{i \cdot t \cdot d\Gamma A}$ is a homomorphism on $\Gamma_0\mathcal{H}$, we get

$$e^{i \cdot t \cdot d\Gamma A}(f \cdot g) = e^{i \cdot t \cdot d\Gamma A}(f) \cdot e^{i \cdot t \cdot d\Gamma A}(g) .$$

By differentiating this identity in $t=0$ we obtain

$$d\Gamma A(f\cdot g) = d\Gamma A(f)\cdot g + f\cdot d\Gamma A(g) .$$

■

Actually the domain $\Gamma_0\mathcal{H} \cap \mathcal{D}(d\Gamma A)$ is dense in $\Gamma\mathcal{H}$. Let $x\in\mathcal{D}(A)$ and fix an $n\in\mathbb{N}$. We then get

$$\frac{1}{t}\cdot(\mathcal{U}_t(x^n) - x^n) = \frac{1}{t}\cdot((U_t x^n) - x^n) = \frac{U_t x - x}{t}\cdot\sum_{k=1}^{n}(U_t x)^{n-k}\cdot x^{k-1} .$$

As $x\in\mathcal{D}(A)$, we will have that

$$\lim_{t\to 0}\frac{U_t\, x - x}{t} = i\cdot Ax ,$$

and according to proposition 1.10B

$$\lim_{t\to 0}(U_t x)^{n-k} = x^{n-k} .$$

This implies that

$$\lim_{t\to 0}\frac{1}{t}\cdot(\mathcal{U}_t(x^n) - x^n) = i\cdot Ax\cdot\sum_{k=1}^{n}x^{n-1} = n\cdot x^{n-1}\cdot(i\cdot Ax) .$$

Hence $x^n\in\mathcal{D}(d\Gamma A)$ for every $x\in\mathcal{D}(A)$ and $n\in\mathbb{N}$, and proposition 1.10B implies that $\mathcal{D}(d\Gamma A)$ is dense in $\Gamma\mathcal{H}$ as $\mathcal{D}(A)$ is dense in $\mathcal{H}$.

Both of the operators ΓA and $d\Gamma A$ are extensions of the operator A, but with different properties. ΓA is multiplicative and $d\Gamma A$ acts according to the Leibniz rule.

Example 2.11: This is a continuation of example 1.19B. The Fourier transformation in $L^2(\mathbb{R}^n)$ is defined as

$$(\mathcal{F}f)(\underline{x}) = (2\pi)^{-\frac{1}{2}n}\cdot\int_{\mathbb{R}^n}\exp(i\cdot\sum_{k=1}^{n}x_k y_k)\cdot f(\underline{y})d\underline{y} ,$$

where $\underline{x} = (x_1, x_2, \ldots, x_n)$ and $\underline{y} = (y_1, y_2, \ldots, y_n)$. It is well known that the Hermite functions form the orthonormal basis of eigenvectors for $\mathcal{F}$.

Let as in lemma 1.14B $f_{|n}$ denote the orthogonal projection of $f \in \Gamma\mathbb{C}^n$ onto the eigenspace of the eigenvalue $(-i)^k$ = the linear span of Hermite functions of order k. Hence in our realization of $\Gamma\mathbb{C}^n$ the operator $\Gamma(-i)$ coincides with the Fourier transformation as we for $f \in \Gamma\mathbb{C}^n$ have that

$$\Gamma(-i)f = \sum_{k=0}^{\infty} (-i)^k \cdot f_{|k} = \mathcal{F}f .$$

Chapter 3: The coherent vectors in $\Gamma\mathcal{H}$

In this chapter we shall consider an important class of vectors, called the coherent vectors, which in physics yields states that minimize the uncertainty in the Heisenberg relation.

Definition 3.1: (cf. [5]) To every $x \in \mathcal{H}$ we assign the coherent vector $\exp x$ which is the sum of the series

$$\exp x = \sum_{n=0}^{\infty} x^n/n! \in \Gamma\mathcal{H} .$$

As the series is absolutely summable, the coherent vector $\exp x$ is well defined.

Proposition 3.2: For every $x,y \in \mathcal{H}$ we have the identities

$$\langle \exp x, \exp y\rangle = e^{\langle x,y\rangle}$$

$$\exp x \cdot \exp y = \exp(x+y) .$$

Proof: We have

$$\langle \exp x, \exp y\rangle = \langle \sum_{n=0}^{\infty} x^n/n!, \sum_{m=0}^{\infty} y^m/m! \rangle = \sum_{n=0}^{\infty} \sum_{m=0}^{\infty} \frac{\langle x^n, y^m\rangle}{n!\cdot m!}$$

$$= \sum_{n=0}^{\infty} \frac{\langle x,y\rangle^n}{n!} = e^{\langle x,y\rangle} ,$$

so that the first identity holds.

We notice, that the other identity requires that the product between two coherent vectors is well defined. We compute

$$\sum_{j+k=n} \frac{1}{j!\cdot k!}\cdot x^j\cdot y^k = \sum_{p=0}^{n} \frac{1}{n!}\cdot\frac{n!}{p!\cdot(n-p)!}\cdot x^p\cdot y^{n-p} = \frac{1}{n!}\cdot(x+y)^n$$

by which we get

$$\sum_{n=0}^{\infty} \Big| \sum_{j+k=n} \frac{1}{j!\cdot k!}\cdot x^j\cdot y^k\Big|^2 = \sum_{n=0}^{\infty} (\tfrac{1}{n!})^2\cdot|(x+y)^n|^2 = e^{|x+y|^2} < \infty .$$

Then

$$\exp x \cdot \exp y = \sum_{n=0}^{\infty} \sum_{j+k=n} \frac{1}{j!\cdot k!}\cdot x^j \cdot y^k = \sum_{n=0}^{\infty} \frac{1}{n!}\cdot (x+y)^n = \exp(x+y)$$

as the terms of the outer series are pairwise orthogonal.

■

Normalized coherent vectors

$$\phi_z = e^{-\frac{1}{2}|z|^2}\cdot \exp z ,$$

define the so-called coherent states, which are of great importance in quantum physics, especially in quantum optics.

<u>**Proposition 3.3:**</u> The mapping

$$\mathcal{H} \ni x \longrightarrow \exp x \in \Gamma\mathcal{H}$$

is uniformly continuous on bounded subsets of $\mathcal{H}$.

<u>**Proof:**</u> Consider $x,y\in\mathcal{H}$ with $|x|,|y| \leq R$, where R is a fixed positive number. By using the estimate in lemma 1.11B we get

$$\begin{aligned}|\exp x - \exp y| &\leq \sum_{n=0}^{\infty} |x^n - y^n|/n! \\ &\leq \sum_{n=0}^{\infty} n!^{\frac{1}{2}}\cdot n\cdot (|x|+|y|)^{n-1}\cdot |x - y|/n! \\ &\leq \sum_{n=0}^{\infty} \frac{n}{\sqrt{n!}}\cdot (2R)^{n-1}\cdot |x - y| \longrightarrow 0 \quad \text{for} \quad x\longrightarrow y .\end{aligned}$$

■

<u>**Theorem 3.4:**</u> For every $f\in\Gamma_0\mathcal{H}$ and $x\in\mathcal{H}$ the product

$$f\cdot \exp x$$

is well defined, and

$$f\cdot \sum_{k=0}^{n} x^k/k! \xrightarrow[n]{} f\cdot \exp x \quad \text{in} \quad \Gamma\mathcal{H} .$$

Proof: Due to proposition 1.3B it is sufficient to consider $f = f_{|m} = a^m$ for $a \in \mathcal{H}$ and $m \in \mathbb{N}$. For

$$g = \exp x = \sum_{n=0}^{\infty} g_n = \sum_{n=0}^{\infty} x^n/n! ,$$

we shall prove that

$$\sum_{n=0}^{\infty} \Big| \sum_{j+k=n} f_{|j} \cdot g_k \Big|^2 < \infty .$$

By using proposition 1.8B we have for every pair $n \geq m$

$$\Big| \sum_{j+k=n} f_{|j} \cdot g_k \Big|^2 = \Big| a^m \cdot \frac{x^{n-m}}{(n-m)!} \Big|^2 \leq \binom{n}{m} |a^m|^2 \cdot \Big| \frac{x^{n-m}}{(n-m)!} \Big|^2$$

$$\leq |a|^{2m} \cdot |x|^{2(n-m)} \cdot \frac{n!}{(n-m)!^2} ,$$

which are known to be the terms of an absolutely summable series.

Then we get

$$f \cdot \exp x = \sum_{n=0}^{\infty} \sum_{j+k=n} f_{|j} \cdot g_k = \sum_{n=m}^{\infty} a^m \cdot x^{n-m}/(n-m)! .$$

We have

$$f \cdot \exp x - f \cdot \sum_{k=1}^{n} x^k/k! = \sum_{k=m+n+1}^{\infty} a^m \cdot x^{k-m}/(k-m)! ,$$

and hence

$$\Big| f \cdot \exp x - f \cdot \sum_{k=1}^{n} x^k/k! \Big| \leq \sum_{k=m+n+1}^{\infty} |a^m \cdot x^{k-m}/(k-m)!|$$

$$\leq \sum_{k=m+n+1}^{\infty} \binom{k}{m}^{\frac{1}{2}} \cdot |a^m| \cdot \Big| \frac{x^{k-m}}{(k-m)!} \Big|$$

$$\leq |a|^m \cdot \sum_{k=m+n+1}^{\infty} \frac{\sqrt{k!}}{(k-m)!} \cdot |x|^{k-m} .$$

Since the last series converges, the theorem follows.

■

We shall now broaden the notation already introduced. For arbitrary $f \in \Gamma\mathcal{H}$ we shall denote by $a^+(f)$ the operator of multiplication by $f \in \Gamma\mathcal{H}$; if $a^+(f)$ is densely defined, $a(f)$ will denote the dual operator of $a^+(f)$. As these operators are often closable (for example in the case $f\in\Gamma_0\mathcal{H}$), we will denote the closure by the same symbol.

In particular, from the above proved theorem it follows that the operators $a^+(\exp x)$, $x\in\mathcal{H}$, are densely defined. Hence they admit dual operators $a(\exp x)$ respectively. From proposition 3.9 we conclude that the dual operators $a(\exp x)$ are densely defined and hence the operators $a^+(\exp x)$ are closable.

<u>**Definition 3.5:**</u> We define the extended Bose algebra setting

$$\Gamma_1\mathcal{H} = \text{span}\left\{ f\cdot\exp x \;\middle|\; x\in\mathcal{H} \text{ and } f\in\Gamma_0\mathcal{H} \right\} .$$

It is obvious that $\Gamma_1\mathcal{H}$ becomes an algebra under the multiplication

$$(f\cdot\exp x)\cdot(g\cdot\exp y) = f\cdot g\cdot\exp(x+y)$$

for all $x,y\in\mathcal{H}$ and $f,g\in\Gamma_0\mathcal{H}$. We notice that

$$\Gamma_0\mathcal{H} \subset \Gamma_1\mathcal{H} \subset \Gamma\mathcal{H} .$$

It turns out that in a natural way several results related to $\Gamma_0\mathcal{H}$ can be extended to include the whole $\Gamma_1\mathcal{H}$.

<u>**Lemma 3.6:**</u> For every $x\in\mathcal{H}$ the operator $a(x) = x^*$ is defined on the whole $\Gamma_1\mathcal{H}$, still fulfilling the Leibniz rule.

<u>**Proof:**</u> Let $\mathcal{D}(x)$ denote the domain of the multiplication operator $a^+(x)$. For $g\in\mathcal{D}(x)$, $f\in\Gamma_0\mathcal{H}$ and $z\in\mathcal{H}$, we have

$$\begin{aligned}
<x\cdot g, f\cdot\exp z> &= \sum_{n=0}^{\infty} <x\cdot g, f\cdot z^n/n!> = \sum_{n=0}^{\infty} <g, x^*(f\cdot z^n/n!)> \\
&= \sum_{n=0}^{\infty} <g, x^*(f)\cdot z^n/n!> + \sum_{n=0}^{\infty} <g, f\cdot x^*(z^n/n!)> \\
&= <g, (x^* f)\cdot\exp z> + \sum_{n=0}^{\infty} <g, f\cdot <x,z>\cdot n\cdot z^{n-1}/n!>
\end{aligned}$$

$$\langle x\cdot g, f\cdot \exp z\rangle = \langle g,(x^*f)\cdot \exp z\rangle + \langle g,\langle x,z\rangle\cdot f\cdot \exp z\rangle .$$

For $f = \emptyset$ the above proves that

$$x^*(\exp z) = \langle x,z\rangle \exp z .$$

Hence the lemma is proved.

■

Corollary 3.7: For every $x\in\mathcal{H}$ we have

$$x^* : \Gamma_1\mathcal{H} \longrightarrow \Gamma_1\mathcal{H}$$

and

$$(x^*)^n \exp z = \langle x,z\rangle^n\cdot \exp z \quad \text{for} \quad z\in\mathcal{H} \quad \text{and} \quad n\in\mathbb{N}_0 .$$

Definition 3.8: For $x\in\mathcal{H}$ we define the operator

$$\exp x^* = \sum_{n=0}^{\infty} (x^*)^n/n!$$

on $f\in\Gamma\mathcal{H}$ such that

$$f \in \mathcal{D}(x^{*n}) \quad \text{for every} \quad n\in\mathbb{N}$$

and

$$\sum_{n=0}^{\infty} |x^{*n}(f)|/n! < \infty .$$

Proposition 3.9: Consider $x\in\mathcal{H}$. The operator $\exp x^*$ is an isomorphism of the algebra $\Gamma_0\mathcal{H}$ onto $\Gamma_0\mathcal{H}$, where

$$(\exp x^*)^{-1} = \exp -x^* .$$

The operators $\exp x^*$ and $a(\exp x)$ are identical on $\Gamma_0\mathcal{H}$.

Proof: Since for every $f\in\Gamma_0\mathcal{H}$ the series

$$\sum_{n=0}^{\infty} n!^{-1}\cdot(x^*)^n f$$

reduces to a finite sum, the operator $\exp x^*$ is defined on the whole $\Gamma_0\mathcal{H}$ with values in $\Gamma_0\mathcal{H}$.

Using induction and the Leibniz rule we easily obtain the

identity

$$x^{*n}(f\cdot g) = \sum_{k=0}^{n} \binom{n}{k}\cdot x^{*k}(f)\cdot x^{*(n-k)}(g)$$

for all $n\in\mathbb{N}_0$ and $f,g\in\Gamma_0\mathcal{H}$. By summation over $n\in\mathbb{N}_0$, and since the series is finite, we get

$$\exp x^*(f\cdot g) = \sum_{n=0}^{\infty} x^{*n}(f\cdot g)/n! = \sum_{n=0}^{\infty}\sum_{k=0}^{n} x^{*k}(f)/k!\cdot x^{*(n-k)}(g)/(n-k)!$$

$$= \sum_{n=0}^{\infty} x^{*n}(f)/n! \cdot \sum_{n=0}^{\infty} x^{*n}(g)/n! = (\exp x^*f)\cdot(\exp x^*g) .$$

Using theorem 3.4 we obtain the last statement

$$\langle f\cdot\exp x,g\rangle = \sum_{n=0}^{\infty} \langle f\cdot x^n/n!,g\rangle = \sum_{n=0}^{\infty} \langle f,x^{*n}(g)/n!\rangle = \langle f,\exp x^*g\rangle .$$

Since the identity

$$(\exp x^*)(\exp -x^*) = I$$

is obvious, the proposition holds.

■

From proposition 3.9 we conclude that the operators $a(\exp x)$ are densely defined and hence the operators $a^+(\exp x)$ are closable for all $x\in\mathcal{H}$.

<u>**Proposition 3.10:**</u> For $x,y\in\mathcal{H}$ we have

$$\exp y \in \mathcal{D}(\exp x^*)$$

and

$$\exp x^* \exp y = e^{\langle x,y\rangle}\exp y .$$

<u>**Proof:**</u> By corollary 3.7 we get

$$x^{*n}(\exp y) = \langle x,y\rangle^n \exp y ,$$

by which

$$\exp x^*(\exp y) = \sum_{n=0}^{\infty} x^{*n}(\exp y)/n! = \sum_{n=0}^{\infty} \langle x,y\rangle^n \exp y/n!$$
$$= e^{\langle x,y\rangle}\exp y .$$

■

<u>**Proposition 3.11:**</u> For every pair $x,y\in\mathcal{H}$ we have on $\Gamma_0\mathcal{H}$ the operator identity

$$\exp x^* \; a^+(\exp y) = e^{\langle x,y\rangle} a^+(\exp y) \exp x^* .$$

<u>**Proof:**</u> First we will show that $\exp x^*$ is defined on the whole $\Gamma_1\mathcal{H}$. It is sufficient to consider elements of the form

$$a^m\cdot \exp z$$

with $a,z\in\mathcal{H}$ and $m\in\mathbb{N}$.

By the Leibniz rule and induction we get

$$x^{*n}(a^m\cdot\exp z) = \sum_{k=0}^{n} \binom{n}{k}\cdot x^{*k}(a^m)\cdot x^{*(n-k)}(\exp z) .$$

For $n>m$ this reduces to

$$x^{*n}(a^m\cdot\exp z) = \sum_{k=0}^{m} \binom{n}{k}\cdot x^{*k}(a^m)\cdot \exp z\cdot\langle x,z\rangle^{n-k} ,$$

by which we get the estimate

$$|x^{*n}(a^m\cdot\exp z)| \le \sum_{k=0}^{m} \binom{n}{k}\cdot |x^{*k}(a^m)\cdot\exp z|\cdot|\langle x,z\rangle|^{n-k} .$$

Letting $C = \max\left\{ |x^{*k}(a^m)\cdot\exp z| \;\middle|\; 0\le k\le m \right\}$, we have

$$|x^{*n}(a^m\cdot\exp z)| \le C\cdot\sum_{k=0}^{m}\binom{n}{k}\cdot|\langle x,z\rangle|^{n-k} = C\cdot\sum_{k=0}^{m}\binom{n}{k}\cdot 1^k\cdot|\langle x,z\rangle|^{n-k}$$
$$= C\cdot(1 + |\langle x,z\rangle|)^n ,$$

by which the series

$$\sum_{n>m} \frac{|x^{*n}(a^m\cdot\exp z)|}{n!} \le \sum_{n>m} C\cdot\frac{(1 + |\langle x,z\rangle|)^n}{n!} \le C\cdot e^{1+|\langle x,z\rangle|}$$

converges, so that $\exp x^*$ becomes defined on the whole $\Gamma_1\mathcal{H}$.

It remains to prove the identity

$$\exp x^* \, a^+(\exp y) = e^{<x,y>} a^+(\exp y) \exp x^*$$

on the whole $\Gamma_0\mathcal{H}$, i.e. to show that

$$\exp x^*(a^m \cdot \exp y) = e^{<x,y>} \cdot \exp y \cdot \exp x^*(a^m)$$

for $a \in \mathcal{H}$ and $m \in \mathbb{N}_0$.

Let

$$f_N = a^m \cdot \sum_{n=0}^{N} y^n/n! = \sum_{n=0}^{N} a^m \cdot y^n/n! \quad \in \quad \mathcal{D}(\exp x^*) .$$

If we prove that

(1) $$f_N \xrightarrow[N]{} f = a^m \cdot \exp y \quad \text{and} \quad f \in \mathcal{D}(\exp x^*)$$

and

(2) $\{\exp x^*(f_N)\}_N$ converges to $e^{<x,y>} \cdot \exp x^*(a^m) \cdot \exp y$,

then since $\exp x^*$ is closable we get

$$\exp x^*(f_N) \xrightarrow[N]{} \exp x^*(f) ,$$

and using the assumption (2) we obtain the desired identity.

The assumption (1) is easy to verify, and then it remains to check (2). Since $\exp x^*$ is an isomorphism, we get

$$\exp x^*(f_N) = \exp x^*(a^m) \cdot \sum_{n=0}^{N} (y + <x,y>\emptyset)^n/n! .$$

By setting $t = <x,y> \in \mathbb{C}$ and since $\exp x^*(a^m) \in \Gamma_0\mathcal{H}$, it is sufficient to prove

$$c^p \cdot \sum_{n=0}^{N} (y + <x,y>\emptyset)^n/n! \xrightarrow[N]{} e^t \cdot c^p \cdot \exp y$$

for some $c \in \mathcal{H}$ and some $p \in \mathbb{N}_0$.

It is easy to prove

$$c^p \cdot \sum_{n=0}^{N} \sum_{k=0}^{N} \frac{t^k}{k!} \cdot \frac{y^n}{n!} \xrightarrow[N]{} c^p \cdot e^t \cdot \exp y$$

by which it is sufficient to get

$$\left| c^p \cdot \sum_{n=0}^{N} \sum_{k=0}^{N} \frac{t^k}{k!} \cdot \frac{y^n}{n!} - c^p \cdot \sum_{n=0}^{N} (y + <x,y>\emptyset)^n/n! \right| \xrightarrow[N]{} 0 .$$

We then have to estimate

$$\left|c^p\cdot\sum_{n=0}^{N}\sum_{k=0}^{N}\frac{t^k}{k!}\cdot\frac{y^n}{n!}-c^p\cdot\sum_{n=0}^{N}(y+<x,y>\emptyset)^n/n!\right| =$$

$$=\left|c^p\cdot\sum_{n=0}^{N}\sum_{k=0}^{N}\frac{t^k}{k!}\cdot\frac{y^n}{n!}-c^p\cdot\sum_{n=0}^{N}\sum_{k=0}^{n}\binom{n}{k}\cdot\frac{t^k\cdot y^{n-k}}{n!}\right|$$

$$=\left|\,c^p\cdot\sum_{n=0}^{N}\sum_{k=0}^{N}\frac{t^k}{k!}\cdot\frac{y^n}{n!}\;-c^p\cdot\sum_{n=0}^{N}\sum_{k=0}^{n}\frac{t^k}{k!}\cdot\frac{y^{n-k}}{(n-k)!}\,\right| .$$

By a standard argument letting the indices run through the diagonal and introducing more terms into the sum, we get the estimate

$$\left|c^p\sum_{n=0}^{N}\sum_{k=0}^{N}\frac{t^k}{k!}\cdot\frac{y^n}{n!}-c^p\sum_{n=0}^{N}(y+<x,y>\emptyset)^n/n!\right|\le\sum_{n=N}^{2\cdot N}\sum_{k=0}^{n}\left|c^p\cdot\frac{t^k}{k!}\cdot\frac{y^{n-k}}{(n-k)!}\right| .$$

By using proposition 1.8B we estimate

$$\sum_{n=N}^{2\cdot N}\sum_{k=0}^{n}\left|\,c^p\cdot\frac{t^k}{k!}\cdot\frac{y^{n-k}}{(n-k)!}\,\right|\le\sum_{n=N}^{\infty}\sum_{k=0}^{n}\frac{|t|^k}{k!}\cdot\left|\,c^p\cdot\frac{y^{n-k}}{(n-k)!}\,\right|$$

$$\le\sum_{n=N}^{\infty}\sum_{k=0}^{n}\frac{|t|^k}{k!}\cdot\left[\frac{(n-k+p)!}{p!(n-k)!}\right]^{\frac{1}{2}}\cdot\sqrt{p!}\cdot|c|^p\cdot\sqrt{(n-k)!}\cdot\frac{|y|^{n-k}}{(n-k)!}$$

$$\le\sum_{n=N}^{\infty}\sum_{k=0}^{n}\frac{|t|^k}{k!}\sqrt{(n+p)!}\cdot|c|^p\frac{|y|^{n-k}}{(n-k)!}=|c|^p\sum_{n=N}^{\infty}\frac{\sqrt{(n+p)!}}{n!}\cdot(|t|+|y|)^n ,$$

which is a summable series, and thus the proposition holds.

■

By using proposition 3.11 we easily get some results, which would be difficult to prove directly.

<u>Corollary 3.12:</u> For $x\in\mathcal{H}$ the operator

$$\exp x^* : \Gamma_1\mathcal{H}\longrightarrow\Gamma_1\mathcal{H}$$

is an isomorphism of the algebra $\Gamma_1\mathcal{H}$ onto itself, and for $y\in\mathcal{H}$ we have on $\Gamma_1\mathcal{H}$ the intertwining

$$\exp x^*\, a^+(\exp y) = e^{<x,y>}\cdot a^+(\exp y)\exp x^* .$$

<u>Proof:</u> The statements of the corollary are simple consequences of proposition 3.11, proposition 3.10 and the fact that the generators

of $\Gamma_1\mathcal{H}$ are of the form

$$f\cdot\exp z \quad \text{with} \quad f\in\Gamma_0\mathcal{H} \quad \text{and} \quad z\in\mathcal{H} \ .$$

Let us prove that the operator is multiplicative. First we notice that for $f\in\Gamma_0\mathcal{H}$ and $a\in\mathcal{H}$ we have

$$\exp x^*(f\cdot\exp a) = \exp x^*(\exp a)\cdot\exp x^*(f) \ .$$

Since $\exp x^*$ is multiplicative on $\Gamma_0\mathcal{H}$ we have for $g\in\Gamma_0\mathcal{H}$ and $b\in\mathcal{H}$

$$\begin{aligned}\exp x^*((fg)\cdot\exp(a+b)) &= \exp x^*(fg)\cdot\exp x^*(\exp(a+b))\\ &= \exp x^*(f)\cdot\exp x^*(g)\cdot e^{<x,a+b>}\cdot\exp(a+b)\\ &= \exp x^*(f\cdot\exp a)\cdot\exp x^*(g\cdot\exp b) \ .\end{aligned}$$

■

Corollary 3.13: Consider $x\in\mathcal{H}$. The operators $a^+(\exp x)$ and $\exp x^*$ are dual to each other on $\Gamma_1\mathcal{H}$, i.e. the operators $\exp x^*$ and $a(\exp x)$ are equal on $\Gamma_1\mathcal{H}$.

Proof: As usual we consider elements in $\Gamma_1\mathcal{H}$ of the form

$$f = f_0\cdot\exp a$$
$$g = g_0\cdot\exp b$$

with $a,b\in\mathcal{H}$ and $f_0,g_0\in\Gamma_0\mathcal{H}$. Using theorem 3.4 we compute

$$\begin{aligned}<a^+(\exp x)f,g> = <f\cdot\exp x,g> &= <f_0\cdot\exp(a+x),g_0\cdot\exp b>\\ &= \sum_{n=0}^{\infty} <(a+x)^n/n!\cdot f_0,g_0\cdot\exp b>\\ &= \sum_{n=0}^{\infty} <f_0,(a^*+x^*)^n/n!(g_0\cdot\exp b)>\\ &= <f_0,\sum_{n=0}^{\infty}(a^*+x^*)^n/n!(g_0\cdot\exp b)>\\ &= <f_0,\exp(x^*+a^*)g_0\cdot\exp b>\\ &= <f_0,\exp(x^*+a^*)(g_0)\cdot e^{<x+a,b>}\exp b>\end{aligned}$$

and further, using the identity $\exp(x^*+a^*) = \exp x^*\exp a^*$ on $\Gamma_0\mathcal{H}$, we get

$$\langle a^+(\exp x)f,g\rangle = \langle f_0,\left[\exp a^* \ \exp x^*(g_0)\right]\cdot \exp a^*(e^{\langle x,b\rangle}\exp b)\rangle$$
$$= \langle f_0,\exp a^*\left[\exp x^*(g_0)\cdot e^{\langle x,b\rangle}\exp b\right]\rangle$$
$$= \langle f_0\cdot \exp a,\exp x^*(g_0\cdot \exp b)\rangle \ .$$

■

Notice that a priori the identity

$$\exp(x^*+a^*) = \exp x^* \ \exp a^*$$

was only shown on $\Gamma_0\mathcal{H}$. Now we know that it is valid on the whole $\Gamma_1\mathcal{H}$ as well.

Example 3.14: (cf. [11]) As mentioned earlier the normalized coherent vectors yield states, which minimize the uncertainty. If $x\in\mathcal{H}$ yields a physical one particle state, i.e. $|x| = 1$, we have the momentum operator

$$p = 2^{-\frac{1}{2}}\cdot i\cdot \hbar\cdot (a^+(x) - a(x))$$

and the position

$$q = 2^{-\frac{1}{2}}\cdot (a^+(x) + a(x)) \ ,$$

with $\hbar$ denoting Planck's constant divided by 2π.

We must calculate

$$\Delta p\cdot\Delta q$$

on the coherent states in $\Gamma\mathcal{H}$, where $(\Delta p)^2 = \langle p^2\rangle - \langle p\rangle^2$ denotes the square of the standard deviation. More specific, let

$$\varnothing_z = e^{-\frac{1}{2}|z|^2}\cdot \exp z$$

yield a coherent state. Then we have

$$\langle p\rangle = \langle \varnothing_z,p(\varnothing_z)\rangle \qquad \langle p^2\rangle = \langle \varnothing_z,p^2(\varnothing_z)\rangle$$
$$\langle q\rangle = \langle \varnothing_z,q(\varnothing_z)\rangle \qquad \langle q^2\rangle = \langle \varnothing_z,q^2(\varnothing_z)\rangle.$$

It is not hard to show that

$$\Delta p\cdot\Delta q = \frac{\hbar}{2} \ ,$$

i.e. that the uncertainty is minimized in the Heisenberg relation.

Example 3.15: In the algebra $\Gamma\mathbb{C}^n = L^2(\mathbb{R}^n)$, defined in example 1.19B, one can compute the coherent vector $\exp \underline{\mathbf{z}}$, $\underline{\mathbf{z}} \in \mathbb{C}^n$, as the function

$$(\exp \underline{z})(\cdot) = e^{-\frac{1}{2}\|\underline{z}\|^2} \cdot e^{\sqrt{2}\langle \cdot , \underline{z} \rangle} \cdot \phi(\cdot) : \mathbb{R}^n \longrightarrow \mathbb{C} .$$

Chapter 4: The Wick ordering and the Weyl relations

Definition 4.1: Consider operators $B_1, B_2, \dots, B_n$, each operator B_k, $k=1,2,\dots,n$, of the form either $a^+(b_k)$ or $a(b_k)$ for a $b_k \in \mathcal{H}$. We define the Wick ordering (often called the normal ordering) $:B_1 B_2 \dots B_n:$ of $B_1 B_2 \dots B_n$ setting

$$:B_1 B_2 \dots B_n: = B_{i_1} B_{i_2} \dots B_{i_n} ,$$

where $i_1, i_2, \dots, i_n$ is a permutation of the numbers $1,2,\dots,n$, chosen in such a way that all the operators $a(b_i)$ appear on the right side of the operators $a^+(b_j)$.

We then extend the ordering linearly.

For $x,y \in \mathcal{H}$ we shall often write shortly

$$x+y^*$$

for

$$a^+(x) + a(y) .$$

Example 4.2: By induction it is easy to show that

$$:(x+y^*)^n: = \sum_{k=0}^{n} \binom{n}{k} \cdot a^+(x^k) \cdot a(y^{n-k})$$

(the binomial formula for commuting operators!).

Lemma 4.3: For every pair $x,y \in \mathcal{H}$ and $n \in \mathbb{N}$ we have the commutation rules

1) $$[a(y)^n, a^+(x)] = n \cdot \langle y,x \rangle \cdot a(y)^{n-1}$$

2) $$[:(x+y^*)^n: , a^+(x)] = n \cdot \langle y,x \rangle \cdot :(x+y^*)^{n-1}:$$
$$[:(x+y^*)^n: , a(y)] = - n \cdot \langle y,x \rangle \cdot :(x+y^*)^{n-1}:$$

3) $$[(x+y^*)^n , a^+(x)] = n \cdot \langle y,x \rangle \cdot (x+y^*)^{n-1}$$
$$[(x+y^*)^n , a(y)] = - n \cdot \langle y,x \rangle \cdot (x+y^*)^{n-1}$$

Proof: All identities are easy to verify by using the operator identity

$$[A^n,B] = \sum_{k=1}^{n} A^{n-k}\cdot[A,B]\cdot A^{k-1}$$

that holds for arbitrary operators A and B.

■

Proposition 4.4: For every $x,y\in\mathcal{H}$ and $n\in\mathbb{N}$ we have

1) $$(x+y^*)^n = n!\cdot\sum_{k=0}^{[\frac{1}{2}n]} \frac{(\frac{1}{2}<y,x>)^k}{k!\cdot(n-2k)!} :(x+y^*)^{n-2k}):$$

2) $$:(x+y^*)^n: = n!\cdot\sum_{k=0}^{[\frac{1}{2}n]} \frac{(-\frac{1}{2}<y,x>)^k}{k!\cdot(n-2k)!} (x+y^*)^{n-2k} .$$

Proof: We will prove 1) by induction. The cases $n=0$ and $n=1$ follow at once, where by definition

$$A^0 = I = \text{the identity operator.}$$

Take $n>1$. Then we have

$$(x+y^*)^n = (x+y^*)^{n-1}a^+(x) + (x+y^*)^{n-1}y^* .$$

Assuming n to be odd, we get $[\frac{n-1}{2}] = \frac{n-1}{2}$. From the induction hypothesis and lemma 4.3 we get

$$(x+y^*)^{n-1}a^+(x) = (n-1)!\cdot\sum_{k=0}^{\frac{n-1}{2}} \frac{(\frac{1}{2}<y,x>)^k}{k!\cdot(n-1-2k)!} :(x+y^*)^{n-1-2k}):a^+(x)$$

$$= (n-1)!\cdot\sum_{k=0}^{\frac{n-1}{2}} \frac{(\frac{1}{2}<y,x>)^k}{k!\cdot(n-1-2k)!} a^+(x) :(x+y^*)^{n-1-2k}:$$

$$+ (n-1)!\cdot\sum_{k=0}^{\frac{n-3}{2}} \frac{(\frac{1}{2}<y,x>)^k}{k!\cdot(n-1-2k)!} (n-1-2k)\cdot<y,x>\cdot:(x+y^*)^{n-2-2k}:$$

$$= (n-1)!\cdot\sum_{k=0}^{\frac{n-1}{2}} \frac{(\frac{1}{2}<y,x>)^k}{k!\cdot(n-1-2k)!} a^+(x) :(x+y^*)^{n-1-2k}:$$

$$+ (n-1)!\sum_{i=1}^{\frac{n-1}{2}} \frac{(\frac{1}{2}<y,x>)^{i-1}}{(i-1)!(n+1-2i)!}(n+1-2i)<y,x>:(x+y^*)^{n-2i}:$$

$$= n!\cdot \sum_{k=0}^{\frac{n-1}{2}} \frac{n-2k}{n} \frac{(\frac{1}{2}<y,x>)^k}{k!\cdot(n-2k)!} a^+(x) \ :(x+y^*)^{n-1-2k}:$$

$$+ n! \sum_{i=1}^{\frac{n-1}{2}} \frac{i}{n} \frac{(\frac{1}{2}<y,x>)^{i-1}}{i!\cdot(n-2i)!} <y,x>:(x+y^*)^{n-2i}:$$

$$= n!\cdot \sum_{k=0}^{\frac{n-1}{2}} \frac{n-2k}{n} \frac{(\frac{1}{2}<y,x>)^k}{k!\cdot(n-2k)!} a^+(x) \ :(x+y^*)^{n-1-2k}:$$

$$+ n! \sum_{i=0}^{\frac{n-1}{2}} \frac{2\cdot i}{n} \frac{(\frac{1}{2}<y,x>)^{i}}{i!\cdot(n-2i)!} :(x+y^*)^{n-2i}: .$$

We return to the first identity

$$(x+y^*)^n = (x+y^*)^{n-1} a^+(x) + (x+y^*)^{n-1} y^*$$

$$= n!\cdot \sum_{k=0}^{\frac{n-1}{2}} \frac{n-2k}{n} \frac{(\frac{1}{2}<y,x>)^k}{k!\cdot(n-2k)!} a^+(x) \ :(x+y^*)^{n-1-2k}:$$

$$+ n! \sum_{i=0}^{\frac{n-1}{2}} \frac{2\cdot i}{n} \frac{(\frac{1}{2}<y,x>)^{i}}{i!\cdot(n-2i)!} :(x+y^*)^{n-2i}: + (x+y^*)^{n-1} y^*$$

and by using the induction hypothesis on the last term, we get

$$= n!\cdot \sum_{k=0}^{\frac{n-1}{2}} \frac{n-2k}{n} \frac{(\frac{1}{2}<y,x>)^k}{k!\cdot(n-2k)!} a^+(x) \ :(x+y^*)^{n-1-2k}:$$

$$+ n!\cdot \sum_{i=0}^{\frac{n-1}{2}} \frac{2\cdot i}{n} \frac{(\frac{1}{2}<y,x>)^{i}}{i!\cdot(n-2i)!} :(x+y^*)^{n-2i}:$$

$$+ (n-1)!\cdot \sum_{k=0}^{\frac{n-1}{2}} \frac{(\frac{1}{2}<y,x>)^k}{k!\cdot(n-1-2k)!} :(x+y^*)^{n-1-2k}):$$

$$= n!\cdot \sum_{k=0}^{\frac{n-1}{2}} \frac{n-2k}{n} \frac{(\frac{1}{2}\langle y,x\rangle)^k}{k!\cdot(n-2k)!}\, a^+(x)\ :(x+y^*)^{n-1-2k}:$$

$$+ n!\cdot \sum_{i=0}^{\frac{n-1}{2}} \frac{2\cdot i}{n} \frac{(\frac{1}{2}\langle y,x\rangle)^i}{i!\cdot(n-2i)!}\ :(x+y^*)^{n-2i}:$$

$$+ n!\cdot \sum_{k=0}^{\frac{n-1}{2}} \frac{n-2k}{n} \frac{(\frac{1}{2}\langle y,x\rangle)^k}{k!\cdot(n-2k)!}\ :(x+y^*)^{n-1-2k}:\ y^* .$$

Summing the first and last terms gives

$$= n!\cdot \sum_{k=0}^{\frac{n-1}{2}} \frac{n-2k}{n} \frac{(\frac{1}{2}\langle y,x\rangle)^k}{k!\cdot(n-2k)!}\ :(x+y^*)^{n-2k}:$$

$$+ n!\cdot \sum_{k=0}^{\frac{n-1}{2}} \frac{2\cdot k}{n} \frac{(\frac{1}{2}\langle y,x\rangle)^k}{k!\cdot(n-2k)!}\ :(x+y^*)^{n-2k}:$$

and since $[\frac{1}{2}n] = \frac{1}{2}(n-1)$ for n odd

$$= n!\cdot \sum_{k=0}^{[\frac{1}{2}n]} \frac{(\frac{1}{2}\langle y,x\rangle)^k}{k!\cdot(n-2k)!}\ :(x+y^*)^{n-2k}):\ .$$

The case of even n, which is very much the same, is left to the reader. The proof of the second identity is left to the reader as well.

■

Definition 4.5: For $x,y\in\mathcal{H}$ we define on $\Gamma_0\mathcal{H}$ the operators

$$\exp(x+y^*) = \sum_{n=0}^{\infty} (x+y^*)^n/n!$$

$$:\exp(x+y^*): = \sum_{n=0}^{\infty} :(x+y^*)^n:/n!\ .$$

The following theorem ascertains that the operators are well defined.

Theorem 4.6: (cf. [19,I.16]) For every $x,y\in\mathcal{H}$ the series which define $\exp(x+y^*)$ and $:\exp(x+y^*):$ are absolutely summable when evaluated on elements of $\Gamma_0\mathcal{H}$, and on $\Gamma_0\mathcal{H}$ we have the operator identity

$$\exp(x+y^*) = e^{\frac{1}{2}\langle y,x\rangle} :\exp(x+y^*): .$$

Proof: We consider elements of the form a^m, $a\in\mathcal{H}$, $m\in\mathbb{N}_0$. For $n\geq m$ we have

$$\begin{aligned}
:(x+y^*)^n:(a^m) &= \sum_{k=0}^{n} \binom{n}{k} a^+(x^{n-k}) y^{*k}(a^m) = \sum_{k=0}^{m} \binom{n}{k} a^+(x^{n-k}) y^{*k}(a^m) \\
&= \sum_{k=0}^{m} \binom{n}{k} a^+(x^{n-k}) \frac{m!}{(m-k)!}\cdot \langle y,a\rangle^k \cdot a^{m-k} \\
&= \sum_{k=0}^{m} \binom{m}{k} \frac{n!}{(n-k)!}\cdot \langle y,a\rangle^k \cdot x^{n-k}\cdot a^{m-k} .
\end{aligned}$$

We estimate

$$\begin{aligned}
\sum_{n=m}^{\infty} n!^{-1}\cdot \left| :(x+y^*)^n:(a^m) \right| &\leq \sum_{n=m}^{\infty} \sum_{k=0}^{m} \binom{m}{k}\cdot \left|\frac{\langle y,a\rangle^k}{(n-k)!}\right| \cdot \left| x^{n-k}\cdot a^{m-k}\right| \\
&\leq \sum_{n=m}^{\infty} \sum_{k=0}^{m} \binom{m}{k}\cdot \left|\frac{\langle y,a\rangle^k}{(n-k)!}\right| \cdot \binom{n+m-2k}{n-k}^{\frac{1}{2}} \left|x^{n-k}\right| \cdot \left|a^{m-k}\right| \\
&= \sum_{n=m}^{\infty} \sum_{k=0}^{m} \binom{m}{k} \left|\frac{\langle y,a\rangle^k}{(n-k)!}\right| \binom{n+m-2k}{n-k}^{\frac{1}{2}} \sqrt{(n-k)!}\cdot |x|^{n-k}\cdot \sqrt{(m-k)!}\cdot |a|^{m-k} \\
&= \sum_{n=m}^{\infty} \sum_{k=0}^{m} \binom{m}{k} \left|\frac{\langle y,a\rangle^k}{(n-k)!}\right| \cdot \left[\frac{(n+m-2k)!}{(n-k)!\cdot(m-k)!}\right]^{\frac{1}{2}} \cdot \sqrt{(n-k)!}\cdot \sqrt{(m-k)!}\cdot \\
&\qquad\qquad |x|^{n-k}\cdot |a|^{m-k} \\
&= \sum_{n=m}^{\infty} \sum_{k=0}^{m} \binom{m}{k} \left|\frac{\langle y,a\rangle^k}{(n-k)!}\right| \cdot \sqrt{(n+m-2k)!}\cdot |x|^{n-k}\cdot |a|^{m-k} \\
&= \sum_{k=0}^{m} \left[\binom{m}{k} |\langle y,a\rangle|^k\cdot |a|^{m-k}\cdot \sum_{n=m}^{\infty} \frac{\sqrt{(n+m-2k)!}}{(n-k)!}\cdot |x|^{n-k} \right] < \infty
\end{aligned}$$

for $k=1,2,\dots,m$.

Consider the absolutely summable series

$$e^{\frac{1}{2}<y,x>} = \sum_{n=0}^{\infty} \frac{(\frac{1}{2}<y,x>)^n}{n!} = \sum_{p=0}^{\infty} \epsilon_p \cdot \frac{(\frac{1}{2}<y,x>)^{\frac{1}{2}p}}{(\frac{1}{2}p)!} ,$$

where $\epsilon_{2p} = 1$ and $\epsilon_{2p+1} = 0$ for $p \in \mathbb{N}_0$. By using proposition 4.4 we get

$$\begin{aligned} e^{\frac{1}{2}<y,x>} \cdot :\exp(x+y^*):(a^m) &= \sum_{p=0}^{\infty} \epsilon_p \cdot \frac{(\frac{1}{2}<y,x>)^{\frac{1}{2}p}}{(\frac{1}{2}p)!} \cdot \sum_{n=0}^{\infty} \frac{:(x+y^*)^n:(a^m)}{n!} \\ &= \sum_{n=0}^{\infty} \sum_{p=0}^{\infty} \epsilon_p \cdot \frac{(\frac{1}{2}<y,x>)^{\frac{1}{2}p}}{(\frac{1}{2}p)!} \cdot \frac{:(x+y^*)^n:(a^m)}{n!} \\ &= \sum_{n=0}^{\infty} \sum_{k=0}^{[\frac{1}{2}n]} \frac{(\frac{1}{2}<<y,x)^k}{k!} \cdot \frac{:(x+y^*)^{n-2k}:(a^m)}{(n-2k)!} \\ &= \sum_{n=0}^{\infty} \frac{(x+y^*)^n}{n!}(a^m) = \exp(x+y^*)(a^m) . \end{aligned}$$

■

Proposition 4.7: For $x,y \in \mathcal{H}$ we have the identity

$$:\exp(x+y^*): = a^+(\exp x) \exp y^*$$

on $\Gamma_0\mathcal{H}$.

Proof: For $f \in \Gamma_0\mathcal{H}$ consider the series

$$\begin{aligned} :\exp(x+y^*):(f) &= \sum_{n=0}^{\infty} \frac{:(x+y^*)^n:(f)}{n!} = \sum_{n=0}^{\infty} \sum_{k=0}^{n} \frac{a^+(x)^k}{k!} \cdot \frac{y^{*(n-k)}}{(n-k)!}(f) \\ &= \sum_{n=0}^{\infty} \frac{a^+(x)^n}{n!} \cdot \sum_{n=0}^{\infty} \frac{a(y)^n}{n!}(f) = a^+(\exp x) \exp y^*(f) . \end{aligned}$$

■

Proposition 4.8: (Simple Campbell–Baker–Hausdorff formula) For $x,y \in \mathcal{H}$ we have on $\Gamma_0\mathcal{H}$ the identities

$$\begin{aligned} \exp(x+y^*) &= e^{\frac{1}{2}<y,x>} a^+(\exp x) \exp y^* \\ &= e^{-\frac{1}{2}<y,x>} \exp y^* \, a^+(\exp x) . \end{aligned}$$

Proof: The first identity is an easy consequence of proposition 4.7 and theorem 4.6. Then according to corollary 3.12 we get

$$a^+(\exp x)\ \exp y^* = e^{-<y,x>} \exp y^*\ a^+(\exp x)$$

and the remaining identity follows.

■

The operators

$$\exp(x+y^*)\ ,\ :\exp(x+y^*):$$

are defined only on $\Gamma_0\mathcal{H}$. It is easy to extend the above identities to $\Gamma_1\mathcal{H}$ so that theorem 4.6, proposition 4.7 and proposition 4.8 hold on $\Gamma_1\mathcal{H}$ as well.

To every $e\in\mathcal{H}$ we assign the operator

$$W_e = \exp(e-e^*)\ :\ \Gamma_1\mathcal{H} \longrightarrow \Gamma_1\mathcal{H}\ .$$

By the Campbell–Baker–Hausdorff formula (proposition 4.8) we have

$$W_e = e^{-\frac{1}{2}|e|^2}\cdot a^+(\exp e)\cdot\exp(-e^*)$$

on $\Gamma_1\mathcal{H}$.

Lemma 4.9: For every $e,f\in\mathcal{H}$ we have on $\Gamma_1\mathcal{H}$ the relation

$$W_e\ W_f = e^{-i\cdot\mathrm{Im}(<e,f>)}\cdot W_{e+f}\ .$$

Proof: Using the Campbell–Baker–Hausdorff formula we get

$$W_e\ W_f = \exp(e-e^*)\ \exp(f-f^*)$$

$$= e^{-\frac{1}{2}|e|^2}\cdot a^+(\exp e)\ \exp(-e^*)\ e^{-\frac{1}{2}|f|^2}\cdot a^+(\exp f)\ \exp(-f^*)$$

and by using corollary 3.12 we get

$$= e^{-\frac{1}{2}(|e|^2+|f|^2)}\cdot a^+(\exp e)\cdot e^{-<e,f>}\cdot a^+(\exp f)\cdot\exp{-e^*}\cdot\exp{-f^*}$$

$$= e^{-\frac{1}{2}(|e|^2+|f|^2+2<e,f>)}\cdot a^+(\exp(e+f))\ \exp{-(e+f)^*}$$

$$= e^{-\frac{1}{2}(<e,f>\ -\ <f,e>)}\cdot e^{-\frac{1}{2}|e+f|^2}\cdot a^+(\exp(e+f))\ \exp{-(e+f)^*}$$

$$= e^{-\frac{1}{2}(<e,f>\ -\ <f,e>)}\cdot W_{e+f} = e^{-i\cdot\mathrm{Im}(<e,f>)}\cdot W_{e+f}\ .$$

■

<u>**Lemma 4.10:**</u> For every $e \in \mathcal{H}$ we have on $\Gamma_1\mathcal{H}$

$$W_e^* = W_{-e}$$

$$W_e W_{-e} = I .$$

<u>**Proof:**</u> For $f,g \in \Gamma_1\mathcal{H}$ we have

$$\langle W_e f,g\rangle = \langle e^{-\frac{1}{2}|e|^2} \cdot a^+(\exp e) \exp(-e^*)f,g\rangle = \langle f,W_{-e}g\rangle .$$

The other relation is a consequence of lemma 4.9.

■

It is now clear that the operator W_e is a unitary transformation of $\Gamma_1\mathcal{H}$ onto $\Gamma_1\mathcal{H}$. We extend W_e by continuity to a unitary transformation of $\Gamma\mathcal{H}$ onto $\Gamma\mathcal{H}$. As a corollary to lemma 4.9 we get the Weyl relations.

<u>**Theorem 4.11:**</u> For every $e,f \in \mathcal{H}$ we have on $\Gamma\mathcal{H}$ the Weyl relations

$$W_e \cdot W_f = e^{-i \cdot \mathrm{Im}(\langle e,f\rangle)} \cdot W_{e+f} .$$

<u>**Example 4.12:**</u> Consider a unitary operator U of $\mathcal{H}$ and an element $e \in \mathcal{H}$. Then it is easy to show that the identity

$$\Gamma U \cdot W_e \cdot \Gamma U^* = W_{(Ue)}$$

holds on the whole $\Gamma\mathcal{H}$.

Chapter 5: Some special operators

In this chapter we are primarily concerned with investigating the operator δ_-^* . This operator links the complex and real wave representation introduced in chapters 6 and 7.

A: Conjugation in Hilbert spaces

Definition 5.1A: A conjugation $\overline{}$ in a Hilbert space $\mathcal{H}$ is a conjugate-linear mapping

$$\overline{} : \mathcal{H} \longrightarrow \mathcal{H}$$

fulfilling

1) $\overline{}$ is an involution
2) $\langle\bar{x},y\rangle = \langle\bar{y},x\rangle$ for all $x,y\in\mathcal{H}$.

Just after proving proposition 2.1 we mentioned that the same procedure can be applied to conjugate linear isometries as well. Hence we extend the conjugation $\overline{}$ to a conjugation on the whole $\Gamma\mathcal{H}$. We abandon in this case the standard notation Γ^- as too complicated. The extension is multiplicative on the subalgebras $\Gamma_0\mathcal{H}$ and $\Gamma_1\mathcal{H}$, and holds them invariant.

An element $f\in\Gamma\mathcal{H}$ will be called real if

$$f = \bar{f} .$$

Theorem 5.2A: Consider fixed $r>0$ and a conjugation in the Hilbert space $\mathcal{H}$. Then the set

$$\{ \exp x \mid x\in\mathcal{H} \text{ is real and } |x|<r \}$$

is total in $\Gamma\mathcal{H}$, i.e. the span of the set is dense in $\Gamma\mathcal{H}$.

Proof: Choose a real orthonormal basis $\{e_n\}_{n\in\mathbb{N}}$ in the Hilbert space $\mathcal{H}$, and consider the set

$$M = \{ \exp x \mid x\in\mathcal{H} \text{ is real and } |x|<r \} .$$

Consider $f\in M^{\perp}$. It is sufficient to show that $f = 0$. According to theorem 1.12B f can be written in the form

$$f = \sum_{\underline{r}} a_{\underline{r}}\cdot e^{\underline{r}} .$$

Let $x\in\mathcal{H}$ denote a real element with $|x|<r$ (i.e. $\exp x \in M$) . Using the fact that $\exp x^{*}$ is multiplicative, we get

$$0 = \langle \exp x, f\rangle = \sum_{\underline{r}} a_{\underline{r}}\cdot\langle \exp x, e^{\underline{r}}\rangle = \sum_{\underline{r}} a_{\underline{r}}\cdot\langle \emptyset, \exp x^{*}(e^{\underline{r}})\rangle$$

$$= \sum_{\underline{r}} a_{\underline{r}}\cdot\langle \emptyset, (\langle x,e_1\rangle\emptyset + e_1)^{r_1}\cdot\ldots\cdot(\langle x,e_n\rangle\emptyset + e_n)^{r_n}\rangle$$

$$= \sum_{\underline{r}} a_{\underline{r}}\cdot\langle \emptyset, \langle x,e_1\rangle^{r_1}\cdot\ldots\cdot\langle x,e_n\rangle^{r_n}\cdot\emptyset\rangle = \sum_{\underline{r}} a_{\underline{r}}\cdot\langle x,e_1\rangle^{r_1}\cdot\ldots\cdot\langle x,e_n\rangle^{r_n} .$$

Letting

$$x = s_1e_1 + s_2e_2 + \ldots + s_ne_n ,$$

where $n\in\mathbb{N}$ and $s_1,s_2,\ldots,s_n\in\mathbb{R}$ are such that

$$s_1^2 + s_2^2 + \ldots + s_n^2 < r^2 ,$$

we get by setting $s_k = 0$ for $k>n$

$$0 = \sum_{\underline{r}} a_{\underline{r}}\cdot s_1^{r_1}\cdot\ldots\cdot s_n^{r_n} .$$

Since $(s_1,s_2,\ldots,s_n)$ run over the ball with radius r we conclude that

$$a_{\underline{r}} = 0 \quad \text{for all } \underline{r} ,$$

by which $f = 0$.

■

<u>B: The functors h_{-}^{*} and δ_{-}^{*}</u>

Let $(\mathcal{H},\langle,\rangle)$ denote a Hilbert space with a fixed chosen conjugation $\bar{\ }$ and an orthonormal basis $\{e_n\}_{n=1}^{\infty}$.

Definition 5.1B: We define the mapping

$$h_-^* : \Gamma_1\mathcal{H} \longrightarrow \Gamma_1\mathcal{H}$$

by the series

$$h_-^* = \sum_{n=1}^{\infty} a(e_n)a(\bar{e}_n) = \sum_{n=1}^{\infty} e_n^* \bar{e}_n^* .$$

The lower index $-$ points out that the definition depends on the choice of the conjugation.

We shall need the following properties of the operator h_-^*.

Theorem 5.2B: The operator h_-^* is well defined and does not depend on the choice of the basis in $\mathcal{H}$, and

1) $\Gamma_0\mathcal{H}$ is an invariant space under h_-^*, i.e. $h_-^*(\Gamma_0\mathcal{H}) \subset \Gamma_0\mathcal{H}$.

2) $h_-^*(a^m) = m(m-1)\langle\bar{a},a\rangle\, a^{m-2}$ for $a\in\mathcal{H}$ and $m\in\mathbb{N}_0$.

3) $h_-^*(a^m \exp z) = [m(m-1)\langle\bar{a},a\rangle a^{m-2} + 2m\langle\bar{a},z\rangle a^{m-1} + \langle\bar{z},z\rangle a^m]\exp z$ with $a,z\in\mathcal{H}$ and $m\in\mathbb{N}_0$.

Proof: 1) is evident and it follows from 2) that the operator is well-defined and does not depend on the chosen orthonormal basis. To prove 2) notice that

$$\bar{e}_n^*(a^m) = m\langle\bar{e}_n,a\rangle a^{m-1}$$

which implies

$$e_n^* \bar{e}_n^* (a^m) = m(m-1)\langle e_n,a\rangle\langle\bar{e}_n,a\rangle a^{m-2}$$

and

$$h_-^*(a^m) = \Big(\sum_{n=1}^{\infty} \langle\bar{a},e_n\rangle\langle e_n,a\rangle\Big)m(m-1)a^{m-2} = m(m-1)\langle\bar{a},a\rangle a^{m-2} .$$

3) is similar, thought a little more technical.

■

Theorem 5.3B: For all $x\in\mathcal{H}$ and $m\in\mathbb{N}$ the commutation relation

$$[(\tfrac{1}{2}\cdot h_-^*)^m, a^+(x)] = m(\tfrac{1}{2}h_-^*)^{m-1}\,\bar{x}^*$$

holds on the whole $\Gamma_1\mathcal{H}$.

Proof: Take $m=1$. It is sufficient to consider elements $f=a^k\cdot\exp z$ with $a,z\in\mathcal{H}$ and $k\in\mathbb{N}_0$. We compute,

$$
\begin{aligned}
e_n^*\bar{e}_n^*(xa^k\exp z) &= \langle\bar{x},e_n\rangle\langle e_n,a\rangle\cdot k\cdot a^{k-1}\exp z + \langle\bar{x},e_n\rangle\langle e_n,z\rangle\cdot a^k\exp z\\
&+ \langle\bar{e}_n,a\rangle\langle e_n,x\rangle\cdot k\cdot a^{k-1}\exp z\\
&+ \langle\bar{e}_n,z\rangle\langle e_n,x\rangle\cdot a^k\exp z + a^+(x)e_n^*\bar{e}_n^*(a^k\exp z)
\end{aligned}
$$

which implies

$$
\begin{aligned}
&e_n^*\bar{e}_n^*a^+(x)[a^k\exp z] - a^+(x)e_n^*\bar{e}_n^*[a^k\exp z] =\\
&= ka^{k-1}\exp z\cdot[\langle\bar{x},e_n\rangle\langle e_n,a\rangle + \langle\bar{a},e_n\rangle\langle e_n,x\rangle]\\
&\quad + a^k\exp z\cdot[\langle\bar{x},e_n\rangle\langle e_n,z\rangle + \langle\bar{z},e_n\rangle\langle e_n,x\rangle] .
\end{aligned}
$$

Since $a^+(x)$ is a closed operator we get by summation over $n\in\mathbb{N}$

$$\tfrac{1}{2}(h_-^*a^+(x) - a^+(x)h_-^*)\cdot a^k\exp z = k\langle\bar{x},a\rangle a^{k-1}\exp z + a^k\langle\bar{x},z\rangle\exp z ,$$

which gives

$$[\tfrac{1}{2}h_-^*,a^+(x)](a^k\exp z) = \bar{x}^*(a^k\exp z) .$$

It is easy to check that for arbitrary operators A,B and $m\in\mathbb{N}$ we have

$$[A^m,B] = \sum_{k=1}^{m} A^{m-k}\,[A,B]\,A^{k-1} .$$

Using this identity and the fact that the operators $\bar{x}^*$ and h_-^* commute we get on $\Gamma_1\mathcal{H}$

$$[(\tfrac{1}{2}h_-^*)^k,a^+(x)] = \sum_{j=1}^{k} (\tfrac{1}{2}h_-^*)^{k-j}\,\bar{x}^*\,(\tfrac{1}{2}h_-^*)^{j-1} = k(\tfrac{1}{2}h_-^*)^{k-1}\bar{x}^* .$$

■

Definition 5.4B: (cf. [20] paragraph 2A) We define the transformation

$$\delta_-^* : \Gamma_0\mathcal{H} \longrightarrow \Gamma_0\mathcal{H}$$

by the series

$$\delta_-^* = \sum_{n=0}^{\infty} (-\tfrac{1}{2}h_-^*)^n/n! .$$

The operator δ_-^* can be generalized to an operator δ_L^* (cf. [20]) by taking in place of a conjugation $^-$ a continuous mapping

$$L : \mathcal{H} \longrightarrow \mathcal{H}$$

which is conjugate linear. Then in the case when L is a real-self-adjoint Hilbert-Schmidt strict contraction, the operator δ_L^* has a dual operator $a^+(\delta_L)$, which is densely defined. In physics the elements

$$\delta_L = a^+(\delta_L)\cdot\emptyset$$

when normalized provide the so-called squeezed states; in [20] they are called ultracoherent vectors.

Proposition 5.5B: The operator δ_-^* is well defined and for $a\in\mathcal{H}$ and $m\in\mathbb{N}$ we have

$$\delta_-^*(a^m) = \sum_{n=0}^{[m/2]} (-\tfrac{1}{2})^n \frac{m!}{(m-2n)!} \frac{\langle\bar{a},a\rangle^n}{n!} a^{m-2n}$$

and especially

$$\delta_-^*(\emptyset) = \emptyset$$
$$\delta_-^*(x) = x \quad \text{for all } x\in\mathcal{H} .$$

Proof: It is enough to prove that

$$\delta_-^*(a^m) = \sum_{n=0}^{[m/2]} (-\tfrac{1}{2})^n \frac{m!}{(m-2n)!} \frac{\langle\bar{a},a\rangle^n}{n!} a^{m-2n} .$$

We start by proving that for all $m,n\in\mathbb{N}$

$$(h_-^*)^n a^m = \begin{cases} \dfrac{m!}{(m-2n)!} \langle\bar{a},a\rangle^n a^{m-2n} & \text{for } m\geq 2n \\ 0 & \text{for } m<2n \end{cases} .$$

The case $n=1$ is clear. Take $n>1$ and assume that the result holds for $n-1$. We have

$$(h_-^*)^n a^m = h_-^*[(h_-^*)^{n-1} a^m]$$
$$= h_-^*\left[\begin{cases} \dfrac{m!}{(m-2n+2)!} \langle\bar{a},a\rangle^{n-1} a^{m-2n+2} & \text{for } m\geq 2n+2 \\ 0 & \text{for } m<2n+2 \end{cases} \right]$$

$$(h_-^*)^n a^m = \begin{cases} \frac{m!}{(m-2n+2)!} \langle \bar{a},a\rangle^{n-1} \; h_-^*[\, a^{m-2n+2} \,] & \text{for } m\geq 2n+2 \\ 0 & \text{for } m<2n+2 \end{cases}$$

$$= \begin{cases} \frac{m!}{(m-2n+2)!} \langle \bar{a},a\rangle^{n-1} (m-2n+2)(m-2n+1)\langle \bar{a},a\rangle a^{m-2n} & \text{for } m\geq 2n \\ 0 & \text{for } m<2n \end{cases}$$

which implies

$$(\, h_-^* \,)^n a^m = \begin{cases} \frac{m!}{(m-2n)!} \langle \bar{a},a\rangle^n \; a^{m-2n} & \text{for } m\geq 2n \\ 0 & \text{for } m<2n \end{cases} .$$

By using this result we get

$$\delta_-^*(a^m) = \sum_{n=0}^{\infty} (-\tfrac{1}{2}h_-^*)^n \frac{a^m}{n!} = \sum_{n=0}^{[m/2]} (-\tfrac{1}{2}h_-^*)^n \frac{a^m}{n!}$$
$$= \sum_{n=0}^{[m/2]} (-\tfrac{1}{2})^n \frac{m!}{(m-2n)!} \frac{\langle \bar{a},a\rangle^n}{n!} a^{m-2n} .$$

■

Theorem 5.6B: On $\Gamma_0\mathcal{H}$ we have

$$[\delta_-^*, a^+(x)] = -\, \delta_-^* \, \bar{x}^* \quad \text{for all} \quad x\in\mathcal{H} .$$

Proof: From theorem 5.3B we get

$$[(\tfrac{1}{2}h_-^*)^m, a^+(x)] = m\cdot(\tfrac{1}{2}h_-^*)^{m-1} \, \bar{x}^* \qquad \text{for all} \quad m\in\mathbb{N}_0 .$$

Hence

$$[(-\tfrac{1}{2}h_-^*)^m, a^+(x)] = -m(-\tfrac{1}{2}h_-^*)^{m-1} \, \bar{x}^* .$$

Since $a^+(x)$ and $\bar{x}^*$ are closed operators, we get

$$[\delta_-^*, a^+(x)] = \sum_{m=1}^{\infty} -m \, (-\tfrac{1}{2}h_-^*)^{m-1} \, \bar{x}^* \, /m! = -\delta_-^* \, \bar{x}^* .$$

■

We will extend the operator δ_-^* to the whole $\Gamma_1\mathcal{H}$. Rewrite theorem 5.6B as the intertwining

$$\delta_-^* \, a^+(x) = (a^+(x) - a(\bar{x}))\delta_-^* .$$

For $n\in\mathbb{N}$ this is extended to an intertwining on $\Gamma_0\mathcal{H}$,

$$\delta_-^* \, a^+(x^n) = (a^+(x) - a(\bar{x}))^n \, \delta_-^* .$$

We write briefly

$$\exp_n(x) = \sum_{k=0}^{n} x^k/k! \ ,$$

which gives that

$$\delta_-^* \exp_n(a^+(x)) = \exp_n(a^+(x) - a(\overline{x}))\ \delta_-^* \ .$$

The limit on the right side exists on $\Gamma_0\mathcal{H}$, so that we may define on $\Gamma_0\mathcal{H}$

$$\delta_-^* \exp(a^+(x)) = \lim_n \delta_-^* \exp_n(a^+(x)) = \lim_n \exp_n(a^+(x) - a(\overline{x}))\ \delta_-^* \ .$$

By the simple Campbell–Baker–Hausdorff formula we obtain on $\Gamma_0\mathcal{H}$

$$\delta_-^* \exp a^+(x) = \exp(a^+(x) - a(\overline{x}))\ \delta_-^*$$

$$= \exp(-\tfrac{1}{2}\langle\overline{x},x\rangle)\ a^+(\exp x)\ a(\exp \overline{x})\ \delta_-^* \ .$$

This way δ_-^* extends to the whole $\Gamma_1\mathcal{H}$. It is easy to prove that the commutation relation from theorem 5.6B holds on $\Gamma_1\mathcal{H}$.

C: The ψ-product

The ψ–product, discussed in this paragraph, appears in [20, chapter 4], cf. [10] as well.

In the preceding paragraph we have proved the commutation rule

$$[\delta_-^*, a^+(x)] = -\delta_-^*\ \overline{x}^* \ ,$$

which can be easily rewritten in the form of intertwining

$$a^+(x)\delta_-^* = \delta_-^*(a^+(x) + \overline{x}^*) \ .$$

Definition 5.1C: For $f\in\Gamma_0\mathcal{H}$ we define the operator

$$f\psi : \Gamma_1\mathcal{H} \longrightarrow \Gamma_1\mathcal{H}$$

by induction as follows.

1) The operator $\emptyset\psi$ is the identity.

2) For $x\in\mathcal{H}$ we have $x\psi = a^+(x) + \overline{x}^*$

3) Let $:\cdot:$ denote the Wick ordering. For powers x^n , $x\in\mathcal{H}$ and $n\in\mathbb{N}$, we define

$$x^n\psi = :(a^+(x) + \overline{x}^*)^n:$$

4) We extend the operation linearly to $\Gamma_0\mathcal{H}$.

It is easy to check that the definition of $f\psi$ does not depend on the choice of representation of $f\in\Gamma_0\mathcal{H}$.

The commutation from theorem 5.6B changes to

$$a^+(x)\delta_-^* = \delta_-^*(x\psi)$$

for $x\in\mathcal{H}$.

Lemma 5.2C: For every $f,g\in\Gamma_0\mathcal{H}$ the operators $f\psi$ and $g\psi$ commute.

Proof: By using the fact that the operator δ_-^* is invertible we get

$$\delta_-^*(x\psi)(y\psi) = a^+(x)\delta_-^*(y\psi) = a^+(x)a^+(y)\delta_-^* ,$$

which concludes the proof

■

Lemma 5.3C: For every $f\in\Gamma_0\mathcal{H}$ the domain of the dual operator of $f\psi$ contains $\Gamma_1\mathcal{H}$, and on $\Gamma_1\mathcal{H}$ the dual operator is identical with the operator $\overline{f}\psi$, i.e. we have

$$< f\psi a,b > = < a,\overline{f}\psi b > \quad \text{for all} \quad a,b\in\Gamma_1\mathcal{H} .$$

Proof: It is sufficient to consider elements $f=x^n$ for $x\in\mathcal{H}$ and $n\in\mathbb{N}_0$.

$$\begin{aligned}
< f\psi a,b > &= < x^n\psi a,b > = < :(x+\overline{x}^*)^n:a,b > \\
&= < \sum_{k=0}^{n} \binom{n}{k} a^+(x^k)(\overline{x}^*)^{n-k}a,b > \\
&= < a, \sum_{k=0}^{n} \binom{n}{k} a^+(\overline{x}^{n-k})(x^*)^n b > \\
&= < a,\overline{x}^n\psi b > = < a,\overline{f}\psi b > .
\end{aligned}$$

■

Thus the operator $f\psi$ is closable. We will denote the closure by the same symbol.

Definition 5.4C: In $\Gamma_0\mathcal{H}$ we introduce the commutative ψ-multiplication by defining

$$f\psi g = (f\psi)(g\psi)\emptyset \quad \text{for} \quad f,g\in\Gamma_0\mathcal{H} .$$

Remark 5.5C: The ψ-multiplication is associative and linear in both variables. Since $(f\psi)\emptyset = f$, we have the identity

$$f\psi g = (f\psi)g .$$

Proposition 5.6C: The annihilation operators obey on $\Gamma_0\mathcal{H}$ the Leibniz rule relative to the ψ-multiplication, i.e. for $a\in\mathcal{H}$ we have

$$a^*(f\psi g) = (a^*f)\psi g + f\psi(a^*g) \quad \text{with} \quad f,g\in\Gamma_0\mathcal{H} .$$

Proof: We shall prove that $[a^*,f\psi] = (a^*f)\psi$ on $\Gamma_0\mathcal{H}$. It is sufficient to consider $f = x^n$ for $x\in\mathcal{H}$ and $n\in\mathbb{N}_0$.

$$\begin{aligned}
[a^*,x^n\psi] &= a^*{:}(x + \bar{x}^*)^n{:} - {:}(x + \bar{x}^*)^n{:}a^* \\
&= \sum_{k=0}^{n} \binom{n}{k} a^* a^+(x^k)\,(\bar{x}^*)^{n-k} - {:}(x + \bar{x}^*)^n{:}a^* \\
&= \sum_{k=0}^{n} \binom{n}{k}\left(k\langle a,x\rangle a^+(x^{k-1})(\bar{x}^*)^{n-k} + a^+(x^k)(\bar{x}^*)^{n-k}a^*\right) \\
&\qquad - {:}(x + \bar{x}^*)^n{:}a^* \\
&= \sum_{k=0}^{n} \binom{n}{k} k\langle a,x\rangle a^+(x^{k-1})(\bar{x}^*)^{n-k} \\
&\qquad + \sum_{k=0}^{n} \binom{n}{k} a^+(x^k)(\bar{x}^*)^{n-k}\, a^* - {:}(x + \bar{x}^*)^n{:}a^* \\
&= \sum_{k=1}^{n} \binom{n}{k} k\langle a,x\rangle a^+(x^{k-1})(\bar{x}^*)^{n-k}
\end{aligned}$$

$$[a^*, x^n\psi] = \sum_{i=0}^{n-1} \binom{n}{i+1}(i+1)\langle a,x\rangle a^+(x^i)(\bar{x}^*)^{n-1-i}$$
$$= \langle a,x\rangle\, n :(x + \bar{x}^*)^{n-1}: = (a^* x^n)\psi .$$

■

Theorem 5.7C: The operator δ_-^* is a homomorphism of $\Gamma_0\mathcal{H}$ equipped with ψ–multiplication into $\Gamma_0\mathcal{H}$ with the original multiplication, i.e.

$$\delta_-^*(f\psi g) = (\delta_-^* f)(\delta_-^* g) \quad \text{for all} \quad f,g\in\Gamma_0\mathcal{H} .$$

Proof: We shall prove that $\delta_-^*(x^n\psi) = (\delta_-^* x^n)\delta_-^*$ on $\Gamma_0\mathcal{H}$, where $(\delta_-^* x^n)$ denotes multiplication by the element. The proof goes by induction.

n=0 : Since $\emptyset\psi = a^+(\emptyset)$, we easily get

$$\delta_-^*(\emptyset\psi) = \delta_-^* a^+(\emptyset) = \delta_-^* = (\delta_-^*\emptyset)\delta_-^* .$$

n=1 : By virtue of theorem 5.6B we get

$$\delta_-^*(x\psi) = a^+(x)\delta_-^* = (\delta_-^* x)\delta_-^* .$$

Now the induction follows : An easy calculation shows that

$$x^n\psi = (x\psi)(x^{n-1}\psi) - (n-1)\langle\bar{x},x\rangle(x^{n-2}\psi) ,$$

which implies

$$\delta_-^*(x^n\psi) = \delta_-^*((x\psi)(x^{n-1}\psi)) - (n-1)\langle\bar{x},x\rangle\delta_-^*(x^{n-2}\psi)$$
$$= \delta_-^*(x\psi)\delta_-^*(x^{n-1}\psi) - (n-1)\langle\bar{x},x\rangle(\delta_-^* x^{n-2})\delta_-^*$$
$$= a^+(x)\delta_-^*(x^{n-1}\psi) - \delta_-^*(\bar{x}^* x^{n-1}\psi)\ \delta_-^*$$
$$= a^+(x)(\delta_-^* x^{n-1})\delta_-^* - \delta_-^*(\bar{x}^* x^{n-1}\psi)\ \delta_-^*$$
$$= [a^+(x)(\delta_-^* x^{n-1}) - \delta_-^*(\bar{x}^* x^{n-1}\psi)]\delta_-^* .$$
$$= (a^+(x)(\delta_-^* x^{n-1}) + [\delta_-^*, a^+(x)]x^{n-1})\delta_-^*$$
$$= (\delta_-^* a^+(x)x^{n-1})\delta_-^* = (\delta_-^* x^n)\delta_-^* .$$

■

Though the operators $f\psi$, $f\in\Gamma_0\mathcal{H}$, act on $\Gamma_1\mathcal{H}$, the ψ–multiplication is defined only on $\Gamma_0\mathcal{H}$. We will extend it to the whole $\Gamma_1\mathcal{H}$. Notice that the operator

$$\delta_-^* = \sum_{n=0}^{\infty} (-\tfrac{1}{2}h_-^*)^n/n!$$

is invertible on $\Gamma_1\mathcal{H}$ with the inverse

$$\hat{\delta}_-^* = \sum_{n=0}^{\infty} (+\tfrac{1}{2}h_-^*)^n/n! \ .$$

Definition 5.8C: We extend the ψ–multiplication over $\Gamma_1\mathcal{H}$ setting for every pair $f,g\in\Gamma_1\mathcal{H}$

$$f\psi g = \hat{\delta}_-^*([\delta_-^* f]\cdot[\delta_-^* g]).$$

Remark 5.9C: The ψ–product is still commutative on $\Gamma_1\mathcal{H}$ and the relation

$$\delta_-^*(f\psi g) = (\delta_-^* f)(\delta_-^* g)$$

holds for all $f,g\in\Gamma_1\mathcal{H}$.

Let us examine the ψ–product between coherent vectors.

Proposition 5.10C: For $x,y\in\mathcal{H}$ we have

$$(\exp x)\psi(\exp y) = e^{<\overline{x},y>}\exp(x + y) \ .$$

Proof: For $z\in\mathcal{H}$ we shall calculate $\delta_-^*(\exp z)$. Since

$$h_-^* \exp z = <\overline{z},z>\exp z$$

we get

$$(-\tfrac{1}{2}h_-^*)^n \exp z = (-\tfrac{1}{2}<\overline{z},z>)^n \exp z \ ,$$

thus giving

$$\delta_-^*(\exp z) = \sum_{n=0}^{\infty} (-\tfrac{1}{2}<\overline{z},z>)^n \exp z/n! = e^{-\frac{1}{2}<\overline{z},z>} \exp z \ .$$

Hence we get

$$\delta_-^*(\exp x \,\psi\, \exp y) = (\delta_-^* \exp x)(\delta_-^* \exp y)$$

$$= e^{-\frac{1}{2}<\bar{x},x>} e^{-\frac{1}{2}<\bar{y},y>} \exp(x+y)$$

$$= e^{-\frac{1}{2}(<\bar{x},x> + <\bar{y},y>)} \exp(x+y)$$

$$= e^{-\frac{1}{2}(<\bar{x},x>+<\bar{y},y>)} \cdot e^{+\frac{1}{2}<\overline{x+y},x+y>} \delta_-^*(\exp(x+y))$$

which implies

$$(\exp x)\psi(\exp y) = e^{+\frac{1}{2}(<\bar{x},y> + <\bar{y},x>)} \exp(x+y)$$

$$= e^{<\bar{x},y>} \exp(x+y) \;.$$

■

Since

$$e^{<\bar{x},y>}\exp(x+y) = (\exp x)\, e^{<\bar{x},y>}(\exp y)$$

$$= (\exp x)\, \exp \bar{x}^*(\exp y)$$

$$= :\exp(x+\bar{x}^*):\, \exp y \;,$$

the operator identity

$$(\exp x)\psi = :\exp(x+\bar{x}^*):$$

holds on the coherent vectors in $\Gamma\mathcal{H}$.

This identity can be shown to hold on the whole $\Gamma_1\mathcal{H}$.

Let us define a new inner product $<,>_0$ by the formula

$$<f,g>_0 = <\delta_-^* f, \delta_-^* g>$$

for all $f,g \in \Gamma_0\mathcal{H}$.

In addition we denote by $a_0^+(x) = x\psi$ the creation operator of $x \in \mathcal{H}$ with respect to the ψ–multiplication. The dual operator $a_0(x)$ to $a_0^+(x)$ with respect to $<,>_0$ coincides with the annihilation operator $a(x)$. We have

$$<a(x)f,g>_0 = <\delta_-^* a(x)f, \delta_-^* g> = <a(x)\delta_-^* f, \delta_-^* g>$$

$$= <\delta_-^* f, a^+(x)\delta_-^* g> = <\delta_-^* f, \delta_-^*(a^+(x) + a(\bar{x}))g>$$

$$= <\delta_-^* f, \delta_-^*(x\psi)g> = <\delta_-^* f, \delta_-^* a_0^+(x)g>$$

$$= <f, a_0^+(x)f>_0 \quad \text{for all} \quad f,g \in \Gamma_0\mathcal{H} \;,$$

and the commutation relation

$$[a_o(x), a_o^+(y)] = [a(x), a^+(y) + a(\bar{y})] = [a(x), a^+(y)]$$

$$= \langle x,y\rangle \emptyset = \langle x,y\rangle_o \emptyset \quad \text{for all} \quad x,y \in \mathcal{H} .$$

Hence $(\Gamma_0\mathcal{H}, \langle , \rangle_o)$ equipped with the ψ-multiplication is a new Bose algebra with vacuum $\emptyset$ and base space $\mathcal{H}$. This algebra will be called the ψ-picture of $\Gamma_0\mathcal{H}$.

Likewise we shall call $\Gamma_1\mathcal{H}, \langle , \rangle_o$ the ψ-picture of the extended Bose algebra $\Gamma_1\mathcal{H}, \langle , \rangle$.

Chapter 6: The complex wave representation

In this chapter we will describe the representation, introduced by Segal [16] and Bargmann [1]. Concerning the priority, cf. [4]. The method by which this representation is introduced originates from [20].

Definition 6.1: For $f \in \Gamma\mathcal{H}$ we define the value of f in $z \in \mathcal{H}$ setting

$$f[z] = \langle \exp z, f \rangle .$$

Since the coherent vectors are total in $\Gamma\mathcal{H}$, the representation $f[\cdot]$ of $f \in \Gamma\mathcal{H}$ is injective.

Lemma 6.2: For every pair $f,g \in \Gamma_1\mathcal{H}$ we have that

$$(f \cdot g)[z] = f[z] \cdot g[z]$$

for all $z \in \mathcal{H}$.

Proof: We start by showing that for $f \in \Gamma_1\mathcal{H}$ we have

$$a(f)\exp z = f^{*}\exp z = \overline{f[z]} \cdot \exp z .$$

For $x \in \mathcal{H}$ and $n \in \mathbb{N}_0$ we get, by using that $\exp z^{*}$ is an isomorphism,

$$\begin{aligned} \langle a(f)\exp z, x^n \rangle &= \langle \exp z, f x^n \rangle = \langle \emptyset, (\exp z^{*} f)(\exp z^{*} x^n) \rangle \\ &= \langle \emptyset, (\exp z^{*} f) \cdot \langle z,x\rangle^n \rangle = \langle z,x\rangle^n \langle \exp z, f\rangle \\ &= f[z] \langle \exp z, x^n \rangle = \langle \overline{f[z]} \cdot \exp z, x^n \rangle . \end{aligned}$$

Since the set $\left\{ x^n \;\middle|\; x \in \mathcal{H} \text{ and } n \in \mathbb{N}_0 \right\}$ is total in $\Gamma\mathcal{H}$, we get

$$f^{*}\exp z = \overline{f[z]} \cdot \exp z ,$$

and

$$\begin{aligned} (f \cdot g)[z] &= \langle \exp z, fg \rangle = \langle f^{*}\exp z, g \rangle \\ &= \langle \overline{f[z]} \cdot \exp z, g \rangle = f[z] \cdot \langle \exp z, g \rangle = f[z] \cdot g[z] . \end{aligned}$$

■

Notice that if $f = x^n$ for $x \in \mathcal{H}$ and $n \in \mathbb{N}$, then

$$f[z] = \langle \exp z, x^n \rangle = \langle z, x \rangle^n$$

which means that in this case $f[\cdot]$ is a grade n homogeneous polynomial of the variable $z \in \mathcal{H}$.

Let $\mathcal{K}$ be a k-dimensional complex Hilbert space and let $\{e_i\}_{i=1}^k$ be an orthonormal basis in $\mathcal{K}$. Identifying $\mathcal{K}$ with $\mathbb{C}^k$ by use of the transformation

$$\mathcal{K} \ni z \longrightarrow (z_1, z_2, \ldots, z_k) \in \mathbb{C}^k$$

where $z_i = \langle e_i, z \rangle$, $i=1,2,\ldots k$, we can define the Lebesgue integration over $\mathcal{K}$ as the corresponding Lebesgue integration over $\mathbb{C}^k$ considered as the 2k-dimensional Euclidean space. We have the following lemma.

__Lemma 6.3:__ Consider a finite dimensional subspace $\mathcal{K}$ of $\mathcal{H}$. Let k denote the dimension of $\mathcal{K}$. For all $f, g \in \Gamma_0 \mathcal{K}$ we have

$$\langle f, g \rangle = \pi^{-k} \cdot \int_{\mathbb{C}^k} \overline{f[z]} \cdot g[z] \cdot \exp(-|z|^2) dz ,$$

where $\mathcal{K} \ni z = (z_1, z_2, \ldots, z_k) \in \mathbb{C}^k$

and $z_j = x_j + i \cdot y_j \in \mathbb{C}$ for $j=1,2,\ldots k$

$$|z|^2 = |z_1|^2 + |z_2|^2 + \ldots + |z_k|^2$$

$dz = dx_1 dy_1 dx_2 dy_2 \ldots dx_k dy_k$ indicates the Lebesgue integration over $\mathbb{R}^{2k}$

Since change of orthonormal basis induces a unitary transformation, by the transformation theorem for the Lebesgue integral of $\mathbb{R}^{2k}$ we get that the integration does not depend on the choice of orthonormal basis in $\mathcal{K}$.

__Proof:__ It is sufficient to prove the lemma on the total set

$$\{ e^{\underline{n}} \mid \underline{n} \in \mathbb{N}_0^k \} ,$$

where $\{e_1, e_2, \ldots, e_k\}$ is an orthonormal basis in $\mathcal{K}$ and

$$e^{\underline{n}} = e_1^{n_1} e_2^{n_2} \ldots e_k^{n_k} .$$

It has been proved that

$$< e^{\underline{n}}, e^{\underline{m}} > = \underline{n}! \cdot \delta_{\underline{n},\underline{m}} ,$$

where $\underline{n}! = n_1!\, n_2! \ldots n_k!$ and δ is the Cronecker delta symbol. Using lemma 6.2 we compute

$$e^{\underline{n}}[z] = < \exp z, e^{\underline{n}} > = <z,e_1>^{n_1} \cdot <z,e_2>^{n_2} \cdot \ldots \cdot <z,e_k>^{n_k} ,$$

and

$$\int_{\mathbb{C}^k} \overline{e^{\underline{n}}[z]}\; e^{\underline{m}}[z] \exp(-|z|^2)dz =$$

$$= \int_{\mathbb{C}^k} \overline{<z,e_1>^{n_1} \ldots <z,e_k>^{n_k}} \cdot <z,e_1>^{m_1} \ldots <z,e_k>^{m_k} \cdot \exp(-|z|^2)dz$$

$$= \prod_{i=1}^{k} \int_{\mathbb{C}} \bar{z}^{n_i} \cdot z^{m_i} \cdot \exp(-|z_i|^2) dx_i dy_i .$$

Without loss of generality we assume that $m \geq n$. Since

$$\int_{\mathbb{C}} \bar{z}^n \cdot z^m \cdot \exp(-|z|^2) \cdot dxdy = 2\pi \cdot \begin{cases} 0 & \text{hvis } n \neq m \\ n!/2 & \text{hvis } n = m \end{cases} ,$$

returning to the first identity we get

$$\int_{\mathbb{C}^k} \overline{e^{\underline{n}}[z]}\; e^{\underline{m}}[z] \cdot \exp(-|z|^2)dz = \prod_{i=1}^{k} \pi \cdot r_i! \cdot \delta_{r_i,s_i} = \pi^k \cdot \underline{r}! \cdot \delta_{\underline{r},\underline{s}} ,$$

which concludes the proof.

■

In appendix 2 we have introduced the gaussian measure $\gamma^{\frac{1}{2}}_{\mathcal{H}}$, defined on Hilbert–Schmidt enlargements $\tilde{\mathcal{H}}$ of $\mathcal{H}$. The details of what follows can be found in the appendix.

Consider a continuous linear functional λ defined on a finite dimensional subspace $\mathcal{K}$ of $\mathcal{H}$. Since λ is of the form

$$\lambda = <a,\cdot> : \mathcal{K} \longrightarrow \mathbb{C} \quad \text{where} \quad a \in \mathcal{K} ,$$

we can extend λ to the whole $\mathcal{H}$ using the same expression, i.e. considering the orthogonal projection

$$p_{\mathcal{K}} : \mathcal{H} \longrightarrow \mathcal{K}$$

and setting for $x \in \mathcal{H}$

$$\lambda(x) = \lambda(p_{\mathcal{K}} x) .$$

By the same procedure we extend λ to a continuous linear functional $\tilde{\lambda}$ on $\tilde{\mathcal{H}}$. We choose $\tilde{\mathcal{H}}$ in such a way that the orthogonal projection $p_{\mathcal{K}}$ extends to an orthogonal projection $\tilde{p}_{\mathcal{K}}$,

$$\tilde{p}_{\mathcal{K}} : \tilde{\mathcal{H}} \longrightarrow \mathcal{K}$$

and define

$$\tilde{\lambda} = \langle a, \tilde{p}_{\mathcal{K}} \cdot \rangle : \tilde{\mathcal{H}} \longrightarrow \mathbb{C} .$$

Since $f[\cdot]$ is a polynomial of linear functionals defined on the space $\mathcal{K}$, we extend in this way the function $f[\cdot]$ from $\mathcal{K}$ to $\tilde{\mathcal{H}}$ for every $f \in \Gamma_0 \mathcal{K}$. We denote the extension by $\tilde{f}[\cdot]$.

Observe that $\Gamma_0 \mathcal{H}$ is a union of all $\Gamma_0 \mathcal{K}$, where $\mathcal{K}$ runs over the set of all finite dimensional subspaces of $\mathcal{H}$. Hence we can reformulate lemma 6.3.

<u>**Corollary 6.4:**</u> For all $f,g \in \Gamma_0 \mathcal{H}$ we have

$$\langle f,g \rangle = \int_{\tilde{\mathcal{H}}} \overline{\tilde{f}[z]} \cdot \tilde{g}[z] \ \gamma_{\mathcal{H}}^{\frac{1}{2}}(dz) .$$

<u>**Theorem 6.5:**</u> (cf. [20] formula 1B3) The isometry

$$\bigcup_{\mathcal{K} \subset \mathcal{H}} \left\{ \Gamma_0 \mathcal{K} \mid \dim(\mathcal{K}) < \infty \right\} \ni f \longrightarrow \tilde{f}[\cdot] \in L^2(\gamma_{\mathcal{H}}^{\frac{1}{2}})$$

extends to an isometry

$$\Gamma \mathcal{H} \ni f \longrightarrow \tilde{f}[\cdot] \in L^2(\gamma_{\mathcal{H}}^{\frac{1}{2}}) ,$$

which is called the complex wave representation.

<u>**Proof:**</u> The theorem is an easy consequence of corollary 6.4 and the fact that $\Gamma_0 \mathcal{H}$ is dense in $\Gamma \mathcal{H}$.

■

We cannot expect the isometry to be unitary as the image of $\Gamma \mathcal{H}$ under this isometry is not invariant under complex conjugation. The relation between the standard multiplication in $\Gamma_0\mathcal{H}$ and the usual multiplication of complex polynomials comes directly from lemma 6.2.

For an arbitrary orthonormal basis $\{e_n\}_{n\in\mathbb{N}}$ in $\mathcal{H}$, we can expand $f\in\Gamma\mathcal{H}$ in the basis from theorem 1.12B,

$$f = \sum_{\underline{r}} a_{\underline{r}} \cdot e^{\underline{r}} .$$

By the Parseval identity we get

$$|f|^2 = \left| \sum_{\underline{r}} a_{\underline{r}} \cdot e^{\underline{r}} \right|^2 = \sum_{\underline{r}} \left| a_{\underline{r}} \cdot e^{\underline{r}} \right|^2 = \sum_{\underline{r}} \underline{r}! \cdot |a_{\underline{r}}|^2$$

thus

$$\Gamma\mathcal{H} = \left\{ f = \sum_{\underline{r}} a_{\underline{r}} \cdot e^{\underline{r}} \;\middle|\; \sum_{\underline{r}} \underline{r}! \cdot |a_{\underline{r}}|^2 < \infty \right\} .$$

If we expand $g\in\Gamma\mathcal{H}$ in the same basis,

$$g = \sum_{\underline{s}} a_{\underline{s}} \cdot e^{\underline{s}} ,$$

we get

$$\langle f,g\rangle = \sum_{\underline{r}} \bar{a}_{\underline{r}} \cdot b_{\underline{r}} \cdot \underline{r}! .$$

For $f\in\Gamma\mathcal{H}$ we have by setting $\underline{r} = (r_1, r_2, \dots, r_n)$

$$\overline{f[z]} = \overline{\langle \exp z, f\rangle} = \langle f, \exp z\rangle = \langle \sum_{\underline{r}} a_{\underline{r}} \cdot e^{\underline{r}}, \exp z\rangle = \sum_{\underline{r}} \bar{a}_{\underline{r}} \cdot \langle e^{\underline{r}}, \exp z\rangle$$

$$= \sum_{\underline{r}} \bar{a}_{\underline{r}} \cdot \langle e_1, z\rangle^{r_1} \cdot \langle e_2, z\rangle^{r_2} \cdot \dots \cdot \langle e_n, z\rangle^{r_n}$$

$$= \sum_{\underline{r}} \bar{a}_{\underline{r}} \cdot z_1^{r_1} \cdot z_2^{r_2} \cdot \dots \cdot z_n^{r_n} ,$$

where $z_i = \langle e_i, z\rangle \in\mathbb{C}$ for $i\in\mathbb{N}$. Identifying $\mathcal{H}$ as the <u>weighted</u> ℓ^2-space $\ell^2(\mathbb{C})$, the function $\overline{f[\cdot]}$ can be considered as an entire function on $\mathcal{H}$.

Therefore the elements of the image of $\Gamma\mathcal{H}$ under the complex wave representation are often called the conjugate holomorphic functions on $\mathcal{H}$.

Notice that we extend some continuous functions on $\mathcal{H}$ to continuous functions on enlargements $\tilde{\mathcal{H}}$ of $\mathcal{H}$. It is awkward from the measure-theoretical point of view since we extend functions defined on a measure zero set to a full measure set (cf. appendix 2).

Example 6.6: This is a continuation of example 2.11. It was pointed out earlier that the operator $\Gamma(-i)$ coincides with the traditional Fourier transformation $\mathcal{F}$ in the space $L^2(R^n)$. Considering a function $f \in \Gamma\mathbb{C}^n = L^2(\mathbb{R}^n)$ and then evaluating the Fourier transform of f in $\underline{z}\in\mathbb{C}^n$ we easily find that

$$(\mathcal{F}f)[\underline{z}] = f[i\underline{z}] ,$$

i.e. that in the complex wave representation Fourier transformations amounts to the composition of $f[\cdot]$ with a 90 degrees rotation of the space $\mathbb{C}^n$.

Chapter 7: The real wave representation

This chapter concerns the representation introduced originally by Segal [15]. We apply the technique developed in [20].

Definition 7.1: For every $f \in \Gamma_1\mathcal{H}$ we define the ψ-value $f_{-}(x)$ of f in $x \in \mathcal{H}$ setting

$$f_{-}(x) = (\delta^*_{-}f)[x] = \langle \exp x, \delta^*_{-}f \rangle .$$

Lemma 7.2: For $f, g \in \Gamma_1\mathcal{H}$ we have that

$$(f\psi g)_{-}(x) = f_{-}(x) \cdot g_{-}(x) \quad \text{for all} \quad x \in \mathcal{H} .$$

Proof: From chapter 5 we have that

$$\delta^*_{-}(f\psi g) = (\delta^*_{-}f)(\delta^*_{-}g) ,$$

hence by virtue of lemma 6.2 we get

$$f\psi g_{-}(x) = (\delta^*_{-}(f\psi g))[x] = ((\delta^*_{-}f)(\delta^*_{-}g))[x]$$
$$= (\delta^*_{-}f)[x](\delta^*_{-}g)[x] = f_{-}(x) \cdot g_{-}(x) .$$

■

For arbitrary $a \in \mathcal{H}$ and $f \in \Gamma_1\mathcal{H}$ we define the derivative of $f_{-}(\cdot)$ in direction a setting

$$\partial_a f_{-}(\cdot) = \frac{d}{dt} f_{-}(t \cdot a + \cdot)_{|t=0} .$$

Lemma 7.3: For fixed $f \in \Gamma_1\mathcal{H}$ and $a, b \in \mathcal{H}$ we have

$$\frac{d}{dt} f_{-}(t \cdot a + b) = (a^* f)_{-}(t \cdot a + b)$$

Proof: Since the operators a^* and δ^*_{-} commute, it is sufficient to prove that

$$\frac{d}{dt} \langle \exp(t \cdot a + b), f \rangle = \langle \exp(t \cdot a + b), a^* f \rangle .$$

This can be reduced to proving that

$$\frac{\exp(h \cdot a) - \emptyset}{h} \longrightarrow a \quad \text{in} \quad \Gamma\mathcal{H} \quad \text{for} \quad h \longrightarrow 0$$

since the operator $a^{+}(\exp x)$ is closed. We compute

$$\frac{\exp(h\cdot a)-\emptyset}{h} - a = \frac{\exp(h\cdot a) - \emptyset - h\cdot a}{h} = \frac{1}{h}\left[\sum_{n=0}^{\infty}\frac{(h\cdot a)^{n}}{n!} - \emptyset - h\cdot a\right]$$

$$= \frac{1}{2}\sum_{n=2}^{\infty}\frac{(h\cdot a)^{n}}{n!} = \frac{1}{h}\sum_{k=0}^{\infty}\frac{(h\cdot a)^{k+2}}{(k+2)!}$$

$$= \sum_{k=0}^{\infty}\frac{a^{k+2}\,h^{k}}{(k+2)!}\cdot h \;.$$

We have for $|h| \leq 1$

$$\left|\,\frac{\exp(h\cdot a)-\emptyset}{h} - a\,\right| \leq \sum_{k=0}^{\infty}\left|\,\frac{a^{k+2}\,h^{k}}{(k+2)!}\cdot h\,\right|$$

$$\leq \sum_{k=0}^{\infty}(k+2)!^{\frac{1}{2}}\,\frac{|a|^{k+2}\,|h|^{k}}{(k+2)!}\cdot|h|$$

i.e.

$$\left|\,\frac{\exp(h\cdot a)-\emptyset}{h} - a\,\right| \leq \sum_{k=0}^{\infty}\frac{|a|^{k+2}}{\sqrt{(k+2)!}}\cdot|h| \;,$$

which – since the series $\sum_{k=0}^{\infty}\frac{|a|^{k+2}}{\sqrt{(k+2)!}}$ is summable – converges to 0 as h tends to 0.

■

Notice that the above argument works not only for real t but also for complex t. For $f\in\Gamma_{1}\mathcal{H}$ the function

$$\mathbb{C} \ni t \longrightarrow f\_(t\cdot a + b) \in \mathbb{C}$$

is actually holomorphic.

The above definition together with the proved lemma gives the identity

$$\partial_{a}f\_(\cdot) = (a^{*}f)\_(\cdot) \;.$$

We introduce the following notation.

Given an arbitrary subspace $\mathcal{K}$ of $\mathcal{H}$ invariant under the conjugation $^{-}$, we denote by $\mathcal{K}$ the real part of $\mathcal{K}$ relative to the conjugation.

In particular H denotes the real part of $\mathcal{H}$.

Lemma 7.4: Let $\mathcal{K}$ denote a finite dimensional subspace of the Hilbert space $\mathcal{H}$, invariant under conjugation $^{-}$. Then for $f,g\in\Gamma_0\mathcal{K} \subset \Gamma_0\mathcal{H}$ we have

$$<f,g> = (2\pi)^{-\frac{1}{2}k}\cdot\int_K \overline{f\_(x)}\cdot g\_(x)\cdot\exp(-\tfrac{1}{2}|x|^2)dx ,$$

where k is the dimension of the real part K of $\mathcal{K}$ and dx indicates the integration with respect to the normalized Lebesgue measure on K.

Proof: For $f,g\in\Gamma_0\mathcal{K}$ we define

$$\langle\langle f,g\rangle\rangle = (2\pi)^{-\frac{1}{2}k}\cdot\int_K \overline{f\_(x)}\cdot g\_(x)\cdot\exp(-\tfrac{1}{2}|x|^2)dx .$$

Choosing an orthonormal basis $\{e_i\}_{i=1}^k$ in K and writing x_j for $<e_j,x>$, we get

$$\langle\langle f,g\rangle\rangle = (2\pi)^{-\frac{1}{2}k}\int_{\mathbb{R}^k} \overline{f\_(x)}\cdot g\_(x)\cdot\exp(-\tfrac{1}{2}|x|^2)\cdot dx_1\ldots dx_k .$$

We will show that $<f,g> = \langle\langle f,g\rangle\rangle$ for all $f,g\in\Gamma_0\mathcal{K}$. It is easy to prove that

$$\langle\langle \emptyset,\emptyset\rangle\rangle = 1 .$$

Furthermore, for $x = \sum_{i=1}^{k} x_i e_i \in K$ we have

$$e_{i-}(x) = <\exp x,\delta_-^* e_i> = <\exp x,e_i> = <x,e_i> = x_i\in\mathbb{R} ,$$

and then we get

$$\begin{aligned}\langle\langle e_i,e_j\rangle\rangle &= (2\pi)^{-\frac{1}{2}k}\int_{\mathbb{R}^k} \overline{e_{i-}(x)}\cdot e_{j-}(x)\cdot\exp(-\tfrac{1}{2}|x|^2)\cdot dx \\ &= (2\pi)^{-\frac{1}{2}k}\int_{\mathbb{R}^k} x_i\cdot x_j\cdot\exp(-\tfrac{1}{2}[x_1^2+\ldots+x_k^2])\cdot dx_1\ldots dx_k = \delta_{ij} ,\end{aligned}$$

so that $<f,g> = \langle\langle f,g\rangle\rangle$ for $f,g\in\mathcal{K}$. To conclude the proof it is sufficient to show that

$$\langle\langle b\cdot f,g\rangle\rangle = \langle\langle f,b^* g\rangle\rangle \text{ for all } f,g\in\Gamma_0\mathcal{K} ,$$

where b^* is the dual to the operator of multiplication by $b\in\mathcal{K}$ with respect to the original inner product $<,>$. By using theorem 5.6B we obtain

$$(bf)\_(x) = \langle \exp x, \delta\_^*(bf)\rangle = \langle \exp x, \delta\_^* a^+(b)f\rangle$$
$$= \langle \exp x, (a^+(b)\delta\_^* - \delta\_^* a(\overline{b}))f\rangle$$
$$= \langle \exp x, a^+(b)\delta\_^* f\rangle - \langle \exp x, \delta\_^* a(\overline{b})f\rangle$$
$$= b[x]\cdot(\delta\_^* f)[x] - (\overline{b}^* f)\_(x) = \langle x,b\rangle f\_(x) - (\overline{b}^* f)\_(x) .$$

Since

$$b = \mathrm{Re}(b) + i\cdot \mathrm{Im}(b) ,$$

where $\mathrm{Re}(b) = \dfrac{b+\overline{b}}{2}$ and $\mathrm{Im}(b) = \dfrac{b-\overline{b}}{2\cdot i}$, without loss of generality we can assume that $b \in K$ is real and that $\|b\| = 1$.

We have

$$\langle\!\langle bf,g\rangle\!\rangle = (2\pi)^{-\frac{1}{2}k}\int_{\mathbb{R}^k} \overline{(bf)\_(x)}\cdot g\_(x)\cdot \exp(-\tfrac{1}{2}|x|^2)\cdot dx$$
$$= (2\pi)^{-\frac{1}{2}k}\int_{\mathbb{R}^k} \overline{[\langle x,b\rangle f\_(x) - (b^* f)\_(x)]} g\_(x)\cdot \exp(-\tfrac{1}{2}|x|^2)\cdot dx$$
$$= (2\pi)^{-\frac{1}{2}k}\int_{\mathbb{R}^k} \overline{f\_(x)}\langle b,x\rangle g\_(x)\cdot \exp(-\tfrac{1}{2}|x|^2)\cdot dx$$
$$- (2\pi)^{-\frac{1}{2}k}\int_{\mathbb{R}^k} \overline{(b^* f)\_(x)} g\_(x)\cdot \exp(-\tfrac{1}{2}|x|^2)\cdot dx$$
$$= (2\pi)^{-\frac{1}{2}k}\int_{\mathbb{R}^k} \overline{f\_(x)}\langle b,x\rangle g\_(x)\cdot \exp(-\tfrac{1}{2}|x|^2)\cdot dx - \langle\!\langle b^* f,g\rangle\!\rangle .$$

Choosing the first basis vector $e_1 = b$ we get

$$\langle\!\langle f,b^* g\rangle\!\rangle = (2\pi)^{-\frac{1}{2}k}\int_{\mathbb{R}^k} \overline{f\_(x)}(b^* g)\_(x)\cdot \exp(-\tfrac{1}{2}|x|^2)\cdot dx$$
$$= (2\pi)^{-\frac{1}{2}k}\int_{\mathbb{R}^k} \overline{f\_(x)}\cdot \left[\frac{d}{dt}\, g\_(t\cdot b + x)\right]_{t=0} \cdot \exp(-\tfrac{1}{2}|x|^2)\cdot dx$$
$$= (2\pi)^{-\frac{1}{2}k}\int_{\mathbb{R}^k} \overline{f\_(x)}\cdot \frac{\partial}{\partial x_1} g\_(x)\cdot \exp(-\tfrac{1}{2}|x|^2) dx_1 \ldots dx_k .$$

Since

$$\int_{\mathbb{R}} \overline{f\_(x)}\cdot \frac{\partial}{\partial x_1} g\_(x)\cdot \exp(-\tfrac{1}{2}|x|^2) dx_1 = - \int_{\mathbb{R}} \frac{\partial}{\partial x_1}\left[\overline{f\_(x)}\cdot \exp(-\tfrac{1}{2}|x|^2)\right] g\_(x) dx_1$$

and

$$\int_{\mathbb{R}} \overline{f\_(x)}\cdot \frac{\partial}{\partial x_1} g\_(x)\cdot \exp(-\tfrac{1}{2}|x|^2) dx_1 =$$
$$= - \int_{\mathbb{R}} \frac{\partial}{\partial x_1}\left[\overline{f\_(x)}\cdot \exp(-\tfrac{1}{2}|x|^2)\right] g\_(x) dx_1$$

$$= -\int_{\mathbb{R}} \left[\overline{\frac{\partial}{\partial x_1} f\_(x)}\cdot \exp(-\tfrac{1}{2}|x|^2) - x_1\overline{f\_(x)}\cdot\exp(-\tfrac{1}{2}|x|^2)\right] g\_(x)dx_1$$

$$= \int_{\mathbb{R}} x_1\overline{f\_(x)}\cdot g\_(x)\cdot\exp(-\tfrac{1}{2}|x|^2)dx_1$$

$$- \int_{\mathbb{R}} \overline{\frac{\partial}{\partial x_1} f\_(x)}\cdot g\_(x)\cdot\exp(-\tfrac{1}{2}|x|^2)dx_1 \ ,$$

returning to the integration over $\mathbb{R}^k$ we get

$$\langle\langle f,b^*g\rangle\rangle = (2\pi)^{-\frac{1}{2}k}\int_{\mathbb{R}^k} \overline{f\_(x)}\cdot\frac{\partial}{\partial x_1} g\_(x)\cdot\exp(-\tfrac{1}{2}|x|^2)dx$$

$$= (2\pi)^{-\frac{1}{2}k}\int_{\mathbb{R}^k} x_1\overline{f\_(x)}\cdot g\_(x)\cdot\exp(-\tfrac{1}{2}|x|^2)dx$$

$$- (2\pi)^{-\frac{1}{2}k}\int_{\mathbb{R}^k} \overline{\frac{\partial}{\partial x_1} f\_(x)}\cdot g\_(x)\cdot\exp(-\tfrac{1}{2}|x|^2)dx$$

$$= (2\pi)^{-\frac{1}{2}k}\int_{\mathbb{R}^k} \langle e_1,x\rangle\overline{f\_(x)}\cdot g\_(x)\cdot\exp(-\tfrac{1}{2}|x|^2)dx$$

$$- (2\pi)^{-\frac{1}{2}k}\int_{\mathbb{R}^k} \overline{(e_1^*f)\_(x)}\cdot g\_(x)\cdot\exp(-\tfrac{1}{2}|x|^2)dx$$

$$= (2\pi)^{-\frac{1}{2}k}\int_{\mathbb{R}^k} \langle b,x\rangle\overline{f\_(x)}\cdot g\_(x)\cdot\exp(-\tfrac{1}{2}|x|^2)dx$$

$$- (2\pi)^{-\frac{1}{2}k}\int_{\mathbb{R}^k} \overline{(b^*f)\_(x)}\cdot g\_(x)\cdot\exp(-\tfrac{1}{2}|x|^2)dx$$

$$= (2\pi)^{-\frac{1}{2}k}\int_{\mathbb{R}^k} \langle b,x\rangle\overline{f\_(x)}\cdot g\_(x)\cdot\exp(-\tfrac{1}{2}|x|^2)dx - \langle\langle b^*f,g\rangle\rangle$$

$$= \langle\langle bf,g\rangle\rangle \ .$$

■

We introduce the gaussian measure $\gamma_H = \gamma_H^1$, sitting on Hilbert–Schmidt enlargements $\tilde{H}$ of the real Hilbert space H, as it is done in appendix 2.

In a similar way as in chapter 6 we extend continuous linear real functionals defined on finite dimensional subspaces of H, to continuous functionals on $\tilde{H}$.

If $\mathcal{K}$ is a finite dimensional subspace of $\mathcal{H}$, invariant under the conjugation $\bar{\ }$, and if K denotes its real part, then we have to extend functionals of the form

$$<z,\cdot> : \mathcal{K} \longrightarrow \mathbb{C} \quad \text{where} \quad z \in \mathcal{K}$$

to $\tilde{H}$.

By setting $x = \mathrm{Re}(z) \in K$ and $y = \mathrm{Im}(z) \in K$, we have

$$z = x+i\cdot y \quad \text{and} \quad <z,\cdot> = <x,\cdot> + i<y,\cdot> .$$

We choose $\tilde{H}$ in such a way that the orthogonal projection p_K of H onto K extends to an orthogonal projection $\tilde{p}_K$ of $\tilde{H}$ onto K . Then we define

$$<z,\cdot>^{\sim} = <x,\tilde{p}_K\cdot> + i<y,\tilde{p}_K\cdot> ,$$

cf. appendix 2.

The extensions will be marked by $\sim$. Then we reformulate lemma 7.4.

Corollary 7.5: Let $\mathcal{K}$ denote a finite dimensional subspace of $\mathcal{H}$, invariant under the conjugation $^{-}$. For every pair $f,g \in \Gamma_0\mathcal{K} \subset \Gamma_0\mathcal{H}$ we have that

$$<f,g> = \int_{\tilde{H}} \overline{\tilde{f}\_(x)}\cdot\tilde{g}\_(x) \ \gamma_H(dx) .$$

Theorem 7.6: (cf. [20] Theorem 4C1) The isometry

$$\bigcup_{\mathcal{K}\subset\mathcal{H}} \left\{ \Gamma_0\mathcal{K} \;\middle|\; \begin{array}{l} \dim(\mathcal{K}) < \infty \\ \mathcal{K} \text{ invariant under } \_ \end{array} \right\} \ni f \longrightarrow \tilde{f}\_(\cdot) \in L^2(\gamma_H)$$

can be extended to a unitary transformation

$$\Gamma\mathcal{H} \ni f \longrightarrow \tilde{f}\_(\cdot) \in L^2(\gamma_H) ,$$

which we shall call the real wave representation.

Proof: The extension to an isometry is trivial. It remains to prove that the extension is surjective. This we do by looking at $\tilde{H}$ within $\mathbb{R}^\infty$.

Since cylinder sets form a base for the Borel sets in $\mathbb{R}^\infty$, functions depending on a finite number of variables will be dense in

$L^2(\gamma_H)$. Hence it is sufficient to observe that the set of polynomials in n variables, $n \in \mathbb{N}$, is dense in $L^2(\mathbb{R}^n, \gamma)$, where γ is the measure

$$\gamma(d\underline{\mathbf{x}}) = (2\pi)^{-\frac{1}{2}n} \cdot \exp(-\tfrac{1}{2}(x_1^2+..+x_n^2))dx_1..dx_n \ .$$

But this is a well known fact and the theorem follows.

∎

We have identified the space $\Gamma\mathcal{H}$ as an L^2–space over the gaussian measure, and in this representation the space $\Gamma_0\mathcal{H}$ corresponds to the space of real variable polynomials. The pointwise multiplication of polynomials in the L^2–space corresponds to the ψ–multiplication in $\Gamma_0\mathcal{H}$ so that the representation does not turn the original multiplication in $\Gamma_0\mathcal{H}$ into the pointwise multiplication as in the case of the complex wave representation.

Chapter 8: Bose algebras of operators

The results of this and the next chapter have their origin in a book of Louisell [11]. Some of those results was primarily intended for preliminaries of [21]. Since they fit so well to this set of lecture notes, they are presented here instead, following the suggestions of the author of [21].

Our goal is a description of a special type of operators, which are compositions of the creation and annihilation operators in the normal form. The operators are linearly generated by operators of the form

$$a^{+}(a)a(f) = af^{*} \quad \text{for} \quad a,f \in \Gamma_0\mathcal{H} .$$

Let I denote the identity operator. We introduce the space of operators

$$\mathcal{O}(\Gamma_0\mathcal{H}) = \left\{ P=\lambda I + \sum_{k=1}^{n} a_k f_k^{*} \;\middle|\; a_k, f_k \in \Gamma_0\mathcal{H} \quad \text{for} \quad k=1,..,n\in\mathbb{N} \quad \text{and} \quad \lambda\in\mathbb{C} \right\} .$$

These operators are defined on the whole $\Gamma_1\mathcal{H}$ with values in $\Gamma_1\mathcal{H}$ and with $\Gamma_0\mathcal{H}$ as an invariant subspace.

We make out of $\mathcal{O}(\Gamma_0\mathcal{H})$ an algebra by introducing the Wick product $:PQ:$ of $P,Q \in \mathcal{O}(\Gamma_0\mathcal{H})$ where PQ is the usual composition of P and Q and $:\cdot:$ denotes the Wick ordering from chapter 4.

The multiplication between the generators becomes

$$:(af^{*})(bg^{*}): = (ab)(fg)^{*} \quad \text{for} \quad a,b,f,g\in\Gamma_0\mathcal{H} .$$

Moreover we define the Hilbert space

$$\mathcal{H}+\mathcal{H}^{*} = \left\{ a^{+}(x) + a(y) = x+y^{*} \;\middle|\; x,y\in\mathcal{H} \right\}$$

with inner product

$$\langle\langle x+y^{*}, u+v^{*} \rangle\rangle = \langle x,u\rangle + \overline{\langle y,v\rangle} \quad \text{for} \quad x,y,u,v\in\mathcal{H} .$$

It is clear that

$$\mathcal{H}+\mathcal{H}^{*} \subset \mathcal{O}(\Gamma_0\mathcal{H}) .$$

Definition 8.1: Let $\gamma_{\mathcal{H}}^{\frac{1}{2}}(dz)$ denote the gaussian measure from chapter 6, defined on Hilbert–Schmidt enlargements $\tilde{\mathcal{H}}$ of $\mathcal{H}$. In the algebra $\mathcal{O}(\Gamma_0\mathcal{H})$ we define the inner product

$$\langle\!\langle P,Q\rangle\!\rangle_N = \int_{\tilde{\mathcal{H}}} \tilde{<}\exp(-z)\cdot P\cdot \exp(z),\exp(-z)\cdot Q\cdot \exp(z)>\cdot\gamma_{\mathcal{H}}^{\frac{1}{2}}(dz) \ .$$

The subscript N in $\langle\!\langle , \rangle\!\rangle_N$ refers to the normal ordering (see theorem 8.8).

Take $P = af^*$ and $Q = bg^*$. Since $f^*\exp z = \overline{f[z]}\exp z$, we get

$$<\exp(-z)\cdot P\cdot \exp(z),\exp(-z)\cdot Q\cdot \exp(z)> = f[z]\overline{g[z]}<a,b> \ .$$

Lemma 8.2: For $a,b,f,g\in\Gamma_0\mathcal{H}$ we have that

$$\langle\!\langle af^*,bg^*\rangle\!\rangle_N = <a,b>\overline{<f,g>} \ .$$

Proof: It is an easy consequence of the complex wave representation and the above.

■

In particular, we observe that $\langle\!\langle , \rangle\!\rangle_N$ and $\langle\!\langle , \rangle\!\rangle$ are equal on the subspace $\mathcal{H}+\mathcal{H}^*$.

Lemma 8.3: Let P^* denote the dual operator of $P \in \mathcal{O}(\Gamma_0\mathcal{H})$ within $\Gamma_0\mathcal{H},<,>$. The operation

$$\mathcal{O}(\Gamma_0\mathcal{H}) \ni P \longrightarrow P^* \in \mathcal{O}(\Gamma_0\mathcal{H})$$

is a complex conjugation in $\mathcal{O}(\Gamma_0\mathcal{H}),\langle\!\langle , \rangle\!\rangle_N$, i.e.

$$\langle\!\langle S^*,T\rangle\!\rangle_N = \langle\!\langle T^*,S\rangle\!\rangle_N \quad \text{for all} \quad S,T \in \mathcal{O}(\Gamma_0\mathcal{H}) \ .$$

Proof: A complex conjugation $\overline{}$, as defined in chapter 5, is a conjugate–linear involution fulfilling

$$<\bar{x},y> = <\bar{y},x> \quad \text{for all} \quad x,y \ .$$

This we check on the generators $af^*,bg^* \in \mathcal{O}(\Gamma_0\mathcal{H})$.

$$\langle\!\langle (af^*)^*, bg^* \rangle\!\rangle_N = \langle\!\langle fa^*, bg^* \rangle\!\rangle_N = \langle f,b\rangle \overline{\langle a,g\rangle}$$

$$= \overline{\langle a,g\rangle \overline{\langle b,f\rangle}} = \overline{\langle\!\langle af^*, gb^* \rangle\!\rangle_N} = \langle\!\langle (bg^*)^*, af^* \rangle\!\rangle_N .$$

■

Definition 8.4: For every pair $x,y \in \mathcal{H}$ we define the creation operator

$$a^+(x+y^*) : \mathcal{O}(\Gamma_0\mathcal{H}) \longrightarrow \mathcal{O}(\Gamma_0\mathcal{H})$$

by

$$a^+(x+y^*)P = :(x+y^*)P: \quad \text{for } P \in \mathcal{O}(\Gamma_0\mathcal{H})$$

and the annihilation operator

$$a(x+y^*) : \mathcal{O}(\Gamma_0\mathcal{H}) \longrightarrow \mathcal{O}(\Gamma_0\mathcal{H})$$

as the dual of $a^+(x+y^*)$.

Notice that $a^+(x+y^*)$ is the operator of Wick multiplication by $(x+y^*)$. It is obvious that in the case of $y = 0$, the operators $a^+(x+y^*)$ and $a(x+y^*)$ reduce to the well known operators $a^+(x)$ and $a(x)$, and therefore the notation is merely an extension.

Lemma 8.5: On the generators $af^* \in \mathcal{O}(\Gamma_0\mathcal{H})$ we have that

1) $a^+(x+y^*)\, I = x+y^*$

$a^+(x+y^*)af^* = (x\cdot a)f^* + a(y\cdot f)^*$

2) $a(x+y^*)\, I = 0$

$a(x+y^*)af^* = (x^*a)f^* + a(y^*f)^*$.

Proof: We only verify 2). For the elements $af^*, bg^* \in \mathcal{O}(\Gamma_0\mathcal{H})$ we get by using lemma 8.2

$$\langle\!\langle a^+(x+y^*)af^*, bg^* \rangle\!\rangle_N = \langle\!\langle (xa)f^* + a(yf)^*, bg^* \rangle\!\rangle_N$$

$$= \langle xa,b\rangle \overline{\langle f,g\rangle} + \langle a,b\rangle \overline{\langle yf,g\rangle} = \langle a,x^*b\rangle \overline{\langle f,g\rangle} + \langle a,b\rangle \overline{\langle f,y^*g\rangle}$$

$$= \langle\!\langle af^*, (x^*b)g^* \rangle\!\rangle_N + \langle\!\langle af^*, b(y^*g)^* \rangle\!\rangle_N = \langle\!\langle af^*, (x^*b)g^* + b(y^*g)^* \rangle\!\rangle_N .$$

Then it follows that $a(x+y^*)bg^* = (x^*b)g^* + b(y^*g)^*$.

■

Care should be taken not to mix elements and operators in the expression for $a(x+y^*)$. Particularly the operator $(y^*f)^*$ is not the same as $f^*a^+(y)$. By using the Leibniz rule we get for $a,b\in\Gamma_0\mathcal{H}$

$$\begin{aligned}\langle (y^*f)^*a,b\rangle &= \langle a,(y^*f)b\rangle = \langle a,y^*(bf) - fy^*b\rangle \\ &= \langle a,y^*(bf)\rangle - \langle a,fy^*b\rangle = \langle ay,bf\rangle - \langle f^*a,y^*b\rangle \\ &= \langle f^*a^+(y)a,b\rangle - \langle a^+(y)f^*a,b\rangle = \langle (f^*a^+(y) - a^+(y)f^*)a,b\rangle .\end{aligned}$$

Hence

$$(y^*f)^* = [f^*,a^+(y)] .$$

Lemma 8.6: For $P \in \mathcal{O}(\Gamma_0\mathcal{H})$ and $x\in\mathcal{H}$ we have the intertwinings

$$Pa^+(x) = (a(x^*) + a^+(x))P$$
$$a(x)P = (a(x) + a^+(x))P .$$

Proof: We prove the identities on the generators

$$af^* \in \mathcal{O}(\Gamma_0\mathcal{H}) .$$

$$\begin{aligned}af^*a^+(x) &= a(f^*a^+(x)) = a\cdot((x^*f)^* + a^+(x)f^*) \\ &= a(x^*f)^* + a^+(x)af^* = (a(x^*) + a^+(x))af^*\end{aligned}$$

$$\begin{aligned}a(x)af^* &= x^*af^* = (x^*a)f^* + ax^*f^* = (x^*a)f^* + a(xf)^* \\ &= (a(x) + a^+(x))\cdot af^* .\end{aligned}$$

■

Theorem 8.7: On the algebra $\mathcal{O}(\Gamma_0\mathcal{H})$ we have the commutation relation

$$[a(x+y^*),a^+(u+v^*)] = \langle\langle x+y^*,u+v^*\rangle\rangle_N I .$$

Proof: We compute on generators $af^*,bg^* \in \mathcal{O}(\Gamma_0\mathcal{H})$.

$$\begin{aligned}&[a(x+y^*),a^+(u+v^*)]af^* \\ &= a(x+y^*)\cdot a^+(u+v^*)af^* - a^+(u+v^*)\cdot a(x+y^*)af^* \\ &= a(x+y^*)\left[(u\cdot a)f^* + a(v\cdot f)^*\right] - a^+(u+v^*)\left[(x^*a)f^* + a(y^*f)^*\right] \\ &= a(x+y^*)[(ua)f^*] + a(x+y^*)[a(vf)^*] - \\ &\qquad a^+(u+v^*)[(x^*a)f^*] - a^+(u+v^*)[a(y^*f)^*]\end{aligned}$$

$$= (x^*(ua))f^* + ua(y^*f)^* + (x^*a)(vf)^* + a(y^*(vf))^* -$$
$$(u\cdot x^*a)f^* - (x^*a)(v\cdot f)^* -$$
$$(ua)(y^*f)^* - a(vy^*f)^*$$
$$= (x^*(ua))f^* + a(y^*(vf))^* - (u\cdot x^*a)f^* - a(vy^*f)^*$$
$$= \left[x^*(ua) - u\cdot x^*a \right]f^* + a\left[(y^*(vf))^* - (vy^*f)^* \right]$$
$$=[(x^*u)a]f^* + a[(y^*v)f]^*$$
$$= \langle x,u\rangle af^* + \overline{\langle y,v\rangle} af^* = \langle\!\langle x+y^*,u+v^*\rangle\!\rangle_N\cdot af^* .$$

■

It is obvious that every operator from the algebra $\mathcal{O}(\Gamma_0\mathcal{H})$ is a polynomial in the operators $a^+(x+y^*)$, $x,y\in\mathcal{H}$, and the identity operator I . Hence we have the following theorem.

<u>**Theorem 8.8:**</u> $\mathcal{O}(\Gamma_0\mathcal{H})$ equipped with the Wick multiplication and inner product $\langle\!\langle,\rangle\!\rangle_N$ is a Bose algebra with the vacuum I and the base $\mathcal{H}+\mathcal{H}^*$; the creation operators are $a^+(x+y^*)$ with the dual annihilation operators $a(x+y^*)$.

We denote by $\overline{\mathcal{O}}(\Gamma_0\mathcal{H}),\langle\!\langle,\rangle\!\rangle_N$ the completion of the space $\mathcal{O}(\Gamma_0\mathcal{H}),\langle\!\langle,\rangle\!\rangle_N$.

Consider for $x+y^* \in \mathcal{H}+\mathcal{H}^*$ the coherent vectors

$$:\exp(x+y^*):$$

in the Hilbert space $\overline{\mathcal{O}}(\Gamma_0\mathcal{H})$.

By virtue of proposition 4.7 the identity

$$:\exp(x+y^*): = \sum_{n=0}^{\infty} :(x+y^*)^n:/n! = a^+(\exp x)\exp y^*$$

holds on every element from $\Gamma_1\mathcal{H}$, and hence we get

$$:\exp(x+y^*): = a^+(\exp x)\exp y^* \quad \text{in} \quad \overline{\mathcal{O}}(\Gamma_0\mathcal{H}) .$$

We denote by $\mathcal{O}(\Gamma_1\mathcal{H})$ the extended Bose algebra of $\mathcal{O}(\Gamma_0\mathcal{H})$, i.e. the span of elements

$$:P\cdot\exp(x+y^*):$$

for $P \in \mathcal{O}(\Gamma_0\mathcal{H})$, $x+y^* \in \mathcal{H}+\mathcal{H}^*$. Then we have

$$\mathcal{O}(\Gamma_1\mathcal{H}) = \text{span}\left\{ a^+(a\cdot\exp x)(b\cdot\exp y)^* \;\middle|\; a,b\in\Gamma_0\mathcal{H},\ x,y\in\mathcal{H} \right\}$$
$$= \text{span}\left\{ a^+(f)g^* \;\middle|\; f,g\in\Gamma_1\mathcal{H} \right\} .$$

Our goal is to construct a Bose algebra with a multiplication, called $_§$, and an inner product $\langle\!\langle , \rangle\!\rangle$ such that the ψ-picture of this new algebra is the algebra $\mathcal{O}(\Gamma_0\mathcal{H}),:\cdot:,\langle\!\langle , \rangle\!\rangle_N$.

Let $\{e_n\}_{n\in\mathbb{N}}$ denote an orthonormal basis in $\mathcal{H}$. We define

$$E_n = 2^{-\frac{1}{2}}\cdot(e_n + e_n^*)$$
$$F_n = 2^{-\frac{1}{2}}\cdot(-i)\cdot(e_n - e_n^*) .$$

Here $\left\{E_n, F_n \mid n\in\mathbb{N}\right\}$ is an orthonormal basis in $\mathcal{H}+\mathcal{H}^*$ consisting of elements which are real with respect to the complex conjugation * . Let the symbol § denote the operator δ^*_- relative to the algebra $\mathcal{O}(\Gamma_1\mathcal{H})$ and the conjugation * . By using the real basis written above and the fact that $a(E_n)^2 + a(F_n)^2 = 2\cdot a(e_n)a(e_n^*)$ we get

$$§ = \exp(-\tfrac{1}{2}\sum_{n=1}^{\infty}\left[a(E_n)^2+a(F_n)^2\right]) = \exp(-\sum_{n=1}^{\infty} a(e_n)a(e_n^*)) ,$$

and the inverse operator $§^{-1}$

$$§^{-1} = \exp(+\sum_{n=1}^{\infty} a(e_n)a(e_n^*)) .$$

Then we define the new multiplication $_§$ in $\mathcal{O}(\Gamma_1\mathcal{H})$ setting

$$P_§Q = §\left[:(§^{-1}P)(§^{-1}Q):\right] \quad \text{for} \quad P,Q \in \mathcal{O}(\Gamma_1\mathcal{H})$$

according to theorem 5.7C and definition 5.8C. Moreover we define the inner product $\langle\!\langle , \rangle\!\rangle$ of $P,Q \in \mathcal{O}(\Gamma_1\mathcal{H})$ setting

$$\langle\!\langle P,Q\rangle\!\rangle = \langle\!\langle\, §^{-1}P, §^{-1}Q \,\rangle\!\rangle_N .$$

Since $§^{-1}$ is the identity operator on $\mathcal{H}+\mathcal{H}^*$, the inner product $\langle\!\langle , \rangle\!\rangle$ restricted to $\mathcal{H}+\mathcal{H}^*$ and the original inner product on $\mathcal{H}+\mathcal{H}^*$ are identical.

For $\mathcal{O}(\Gamma_0\mathcal{H})$ as an algebra with multiplication $_§$ and inner product $\langle\!\langle , \rangle\!\rangle$ we use the notation $\Gamma_0(\mathcal{H}+\mathcal{H}^*)$ and write $\Gamma_1(\mathcal{H}+\mathcal{H}^*)$ for the algebra $\mathcal{O}(\Gamma_1\mathcal{H}),_§,\langle\!\langle , \rangle\!\rangle$.

We shall write $\Gamma(\mathcal{H}+\mathcal{H}^*)$ for the completion of the space $\Gamma_0(\mathcal{H}+\mathcal{H}^*),\langle\!\langle,\rangle\!\rangle$.

Theorem 8.9: $\Gamma_0(\mathcal{H}+\mathcal{H}^*),_{\S},\langle\!\langle,\rangle\!\rangle$ is a Bose algebra with the vacuum I , the base $\mathcal{H}+\mathcal{H}^*$, the creation operators

$$\mathcal{A}^+(x+y^*) = a^+(x+y^*) - a(y+x^*)$$

and their dual annihilation operators

$$\mathcal{A}(x+y^*) = a(x+y^*) \ .$$

Proof: We must show the following:

$\mathcal{A}(x+y^*)\ I = 0$

$[\mathcal{A}(x+y^*),\mathcal{A}^+(u+v^*)] = \langle\!\langle x+y^*,u+v^*\rangle\!\rangle\ I$

$\mathcal{A}^+(x+y^*)$ generates from the vacuum I the whole $\Gamma_0(\mathcal{H}+\mathcal{H}^*) = \mathcal{O}(\Gamma_0\mathcal{H})$ and finally

$\mathcal{A}^+(x+y^*)$ and $\mathcal{A}(x+y^*)$ are dual operators on $\Gamma_0(\mathcal{H}+\mathcal{H}^*),\langle\!\langle,\rangle\!\rangle$.

From lemma 8.5 we have that

$$\mathcal{A}(x+y^*)\ I = a(x+y^*)\ I = 0 \ ,$$

and from the commutation relation in theorem 8.7 we have

$$\begin{aligned}[\mathcal{A}(x+y^*),\mathcal{A}^+(u+v^*)] &= [a(x+y^*),a^+(x+y^*) - a(y+x^*)] \\ &= [a(x+y^*),a^+(x+y^*)] = \langle\!\langle x+y^*,u+v^*\rangle\!\rangle_N\cdot I \\ &= \langle\!\langle x+y^*,u+v^*\rangle\!\rangle I \ .\end{aligned}$$

Since $\mathcal{A}^+(x)a = xa$ and $\mathcal{A}^+(y^*)f^* = (yf)^*$ for all $x,y\in\mathcal{H}$ and $a,f\in\Gamma_0\mathcal{H}$, it is easily seen that the creation operators $\mathcal{A}^+(x+y^*)$ and the identity operator I generate the whole algebra $\Gamma_0(\mathcal{H}+\mathcal{H}^*)$.

It remains to prove that $\mathcal{A}^+(x+y^*)$ and $\mathcal{A}(x+y^*)$ are dual operators on $\Gamma_0(\mathcal{H}+\mathcal{H}^*),\langle\!\langle,\rangle\!\rangle$. We need the commutation from theorem 5.6B

$$\delta_-^* a^+(x) = (a^+(x) - a(\bar{x}))\delta_-^* \quad \text{for } x\in\mathcal{H} \ ,$$

which in the algebra $\Gamma_1(\mathcal{H}+\mathcal{H}^*) = \mathcal{O}(\Gamma_1\mathcal{H})$ corresponds to

$$\S\mathcal{A}^+(x+y^*) = (\mathcal{A}^+(x+y^*) - \mathcal{A}(y+x^*))\S \ .$$

By using the fact that § and $\mathcal{A}(x+y^*)$ commute and applying the operator $a^+(x+y^*)$ we get

$$\mathcal{A}^+(x+y^*)\S = \S a^+(x+y^*) \ ,$$

and hence

$$\S^{-1}\mathcal{A}^+(x+y^*) = a^+(x+y^*)\S^{-1} \ .$$

For $P,Q \in \Gamma_0(\mathcal{H}+\mathcal{H}^*)$ and $x+y^* \in \mathcal{H}+\mathcal{H}^*$ we have

$$\begin{aligned}\langle\!\langle \mathcal{A}^+(x+y^*)P,Q\rangle\!\rangle &= \langle\!\langle \S^{-1}\mathcal{A}^+(x+y^*)P,\S^{-1}Q\rangle\!\rangle_N = \langle\!\langle a^+(x+y^*)\S^{-1}P,\S^{-1}Q\rangle\!\rangle_N \\ &= \langle\!\langle \S^{-1}P,a(x+y^*)\S^{-1}Q\rangle\!\rangle_N = \langle\!\langle \S^{-1}P,\mathcal{A}(x+y^*)\S^{-1}Q\rangle\!\rangle_N \\ &= \langle\!\langle \S^{-1}P,\S^{-1}\mathcal{A}(x+y^*)Q\rangle\!\rangle_N = \langle\!\langle P,\mathcal{A}(x+y^*)Q\rangle\!\rangle \ .\end{aligned}$$

■

Definition 8.10: We denote by $\mathcal{P}(\mathcal{H}+\mathcal{H}^*)$ the set of polynomials of a finite number of variables

$$\langle\cdot,a\rangle \quad \text{and} \quad \langle b,\cdot\rangle \quad \text{where} \quad a,b\in\mathcal{H} \ .$$

Then we define the transformation

$$\pi : \Gamma_0(\mathcal{H}+\mathcal{H}^*) = \mathcal{O}(\Gamma_0\mathcal{H}) \longrightarrow \mathcal{P}(\mathcal{H}+\mathcal{H}^*)$$

by

$$(\pi P)[z] = \langle \exp z,(\exp{-z})P(\exp z)\rangle \quad \text{for all} \quad z\in\mathcal{H} \ ,$$

where $P \in \Gamma_0(\mathcal{H}+\mathcal{H}^*)$.

It turns out that $\pi P[\cdot]$ correspond to the Wigner distribution function, cf. [21,25].

Using the generators $P = af^*$ we have that

$$(\pi P)[z] = \langle \exp z,(\exp{-z})af^*(\exp z)\rangle = a[z]\cdot\overline{f[z]} \ .$$

If we extend the functions $\pi P[\cdot] \in \mathcal{P}(\mathcal{H}+\mathcal{H}^*)$ in the usual way from $\mathcal{H}$ to Hilbert–Schmidt enlargements $\tilde{\mathcal{H}}$ and denote the space of extensions by $\tilde{\mathcal{P}}(\mathcal{H}+\mathcal{H}^*)$, we get that

$$\tilde{\mathcal{P}}(\mathcal{H}+\mathcal{H}^*) \subset L^2(\gamma_{\mathcal{H}}^{\frac{1}{2}}) \ ,$$

where $\gamma_{\mathcal{H}}^{\frac{1}{2}}$ is the gaussian measure, defined on $\tilde{\mathcal{H}}$.

Theorem 8.11: For all $P,Q \in \Gamma_0(\mathcal{H}+\mathcal{H}^*)$ we have the identity

$$\langle\!\langle P,Q\rangle\!\rangle = \int_{\tilde{\mathcal{H}}} \pi\overline{P[z]}\cdot\pi Q[z]\cdot\gamma_{\mathcal{H}}^{\frac{1}{2}}(dz) \ ,$$

where for simplicity of notation we have omitted the usual $\sim$. It is then automatically understood that the functions are extended from $\mathcal{H}$ to a Hilbert–Schmidt enlargement $\tilde{\mathcal{H}}$.

Proof: We define

$$\langle\langle P,Q\rangle\rangle' = \int_{\tilde{\mathcal{H}}} \pi\overline{P[z]}\cdot\pi Q[z]\cdot\gamma_{\mathcal{H}}^{\frac{1}{2}}(dz) .$$

Thus for $af^*, bg^* \in \Gamma_0(\mathcal{H}+\mathcal{H}^*)$ we get that

$$\overline{\pi(af^*)[z]}\cdot\pi(bg^*)[z] = \overline{a[z]\overline{f[z]}}\cdot b[z]\overline{g[z]}$$
$$= \overline{a[z]g[z]}\cdot b[z]f[z] = \overline{(ag[z])}\cdot(bf)[z] ,$$

and then by virtue of the complex wave representation we get

$$\langle\langle af^*, bg^*\rangle\rangle' = \langle ag,bf\rangle .$$

Using the elements from the base $\mathcal{H}+\mathcal{H}^*$ we get

$$\langle\langle x+y^*, u+v^*\rangle\rangle' = \langle x,u\rangle + \langle xv,\emptyset\rangle + \langle\emptyset,uy\rangle + \langle v,y\rangle$$
$$= \langle x,u\rangle + \overline{\langle y,v\rangle} = \langle\langle x+y^*, u+v^*\rangle\rangle .$$

It remains to show that $\mathcal{A}^+(x+y^*)$ and $\mathcal{A}(x+y^*)$ are dual operators with respect to $\langle\langle,\rangle\rangle'$. We prove this by using the generators $af^*, bg^* \in \Gamma_0(\mathcal{H}+\mathcal{H}^*)$. For $x+y^*\in\mathcal{H}+\mathcal{H}^*$ we have

$$\langle\langle \mathcal{A}^+(x+y^*)af^*, bg^*\rangle\rangle' = \langle\langle a^+(x+y^*)af^* - a(y+x^*)af^*, bg^*\rangle\rangle'$$
$$= \langle xag,bf\rangle + \langle ag,byf\rangle - \langle (y^*a)g,bf\rangle - \langle ag,b(x^*f)\rangle$$
$$= \langle ag,(x^*b)f\rangle + \langle a(y^*g),bf\rangle$$
$$= \langle\langle af^*,(x^*b)g^*\rangle\rangle' + \langle\langle af^*,b(y^*g)^*\rangle\rangle'$$
$$= \langle\langle af^*,\mathcal{A}(x+y^*)bg^*\rangle\rangle' .$$

■

While proving theorem 8.11 the following corollary was shown:

Corollary 8.12: On the generators $af^*, bg^* \in \Gamma_0(\mathcal{H}+\mathcal{H}^*)$ we have that

$$\langle\langle af^*, bg^*\rangle\rangle = \langle ag,bf\rangle .$$

Lemma 8.13: The operation

$$\Gamma_0(\mathcal{H}+\mathcal{H}^*) \ni P \longrightarrow P^* \in \Gamma_0(\mathcal{H}+\mathcal{H}^*)$$

is a complex conjugation in $(\Gamma_0(\mathcal{H}+\mathcal{H}^*),\langle\langle,\rangle\rangle)$, i.e.

$$\langle\langle S^*,T\rangle\rangle = \langle\langle T^*,S\rangle\rangle \quad \text{for all} \quad S,T \in \Gamma_0(\mathcal{H}+\mathcal{H}^*) .$$

Proof: As usual we check the identity on the generators

$$af^*,bg^* \in \Gamma_0(\mathcal{H}+\mathcal{H}^*) .$$

We have

$$\langle\langle (af^*)^*,bg^*\rangle\rangle = \langle\langle fa^*,bg^*\rangle\rangle = \langle fg,ab\rangle = \overline{\langle ab,fg\rangle}$$

$$= \overline{\langle\langle af^*,(gb^*)^*\rangle\rangle} = \langle\langle (bg^*)^*,af^*\rangle\rangle .$$

■

Notice that the same result concerning the algebra $\mathcal{O}(\Gamma_0\mathcal{H}),\langle\langle,\rangle\rangle_N$ has been proved in lemma 8.3.

From the definition of $\Gamma_0(\mathcal{H}+\mathcal{H}^*)$ we get the following results.

Theorem 8.14: In the Bose algebra $\Gamma_0(\mathcal{H}+\mathcal{H}^*),_{\S},\langle\langle,\rangle\rangle$ with the complex conjugation * the ψ-product and the Wick product are the same, thus

$$P\psi Q = :PQ: \quad \text{for all} \quad P,Q \in \Gamma_0(\mathcal{H}+\mathcal{H}^*) .$$

Theorem 8.15: The Bose algebra $\mathcal{O}(\Gamma_0\mathcal{H}),\langle\langle,\rangle\rangle_N$ is the ψ-picture of the Bose algebra $\Gamma_0(\mathcal{H}+\mathcal{H}^*),\langle\langle,\rangle\rangle$.

Chapter 9: Wave representations of $\Gamma(\mathcal{H}+\mathcal{H}^*)$

In this chapter we shall study some wave representations of $\Gamma(\mathcal{H}+\mathcal{H}^*)$, equipped with the complex conjugation *, discussed in lemma 8.13.

In chapter 8 we derived the expression

$$:\exp(x+y^*): = a^+(\exp x)\exp y^*$$

for the coherent vectors in $\bar{\mathcal{O}}(\Gamma_0\mathcal{H})$.

Let us denote by

$$\exp_{\S}(x+y^*) = \sum_{n=0}^{\infty} (x+y^*)^{\S n}/n!$$

the coherent vectors of $\Gamma(\mathcal{H}+\mathcal{H}^*)$. For $P \in \Gamma_0(\mathcal{H}+\mathcal{H}^*) = \mathcal{O}(\Gamma_0\mathcal{H})$ we have

$$(x+y^*)^{\S n} = \S(:(\S^{-1}(x+y^*))^n:) = \S(:(x+y^*)^n:)$$

and thus

$$\begin{aligned}
\langle\!\langle \exp_{\S}(x+y^*),P\rangle\!\rangle &= \sum_{n=0}^{\infty} \langle\!\langle (x+y^*)^{\S n}/n!,P\rangle\!\rangle \\
&= \sum_{n=0}^{\infty} \langle\!\langle \S(:(x+y^*)^n:/n!),P\rangle\!\rangle \\
&= \sum_{n=0}^{\infty} \langle\!\langle :(x+y^*)^n:/n!,\S^{-1}P\rangle\!\rangle_N \\
&= \langle\!\langle :\exp(x+y^*):,\S^{-1}P\rangle\!\rangle_N = \langle\!\langle \S:\exp(x+y^*):,P\rangle\!\rangle .
\end{aligned}$$

Hence

$$\exp_{\S}(x+y^*) = \S:\exp(x+y^*): .$$

In chapter 8 we computed the reduced expression for the operator $\S$

$$\S = \exp(-\sum_{n=1}^{\infty} a(e_n)a(e_n^*)) ,$$

where $\{e_n\}_{n\in\mathbb{N}}$ is an orthonormal basis in $\mathcal{H}$. Then we get

$$\sum_{n=1}^{\infty} a(e_n)a(e_n^*):\exp(x+y^*): = \sum_{n=1}^{\infty} a(e_n)\langle\!\langle e_n^*, x+y^*\rangle\!\rangle_N:\exp(x+y^*):$$

$$= \sum_{n=1}^{\infty} \langle\!\langle e_n, x+y^*\rangle\!\rangle_N \cdot \langle\!\langle e_n^*, x+y^*\rangle\!\rangle_N:\exp(x+y^*):$$

$$= \sum_{n=1}^{\infty} \langle e_n, x\rangle\overline{\langle e_n, y\rangle}:\exp(x+y^*):$$

$$= \sum_{n=1}^{\infty} \langle e_n, x\rangle\langle y, e_n\rangle:\exp(x+y^*):$$

$$= \langle y, x\rangle:\exp(x+y^*):$$

and hence

$$\exp_{\S}(x+y^*) = e^{-\langle y,x\rangle}:\exp(x+y^*): .$$

For $P \in \Gamma(\mathcal{H}+\mathcal{H}^*)$ the value of P in $x+y^* \in \mathcal{H}+\mathcal{H}^*$ is

$$P[x+y^*] = \langle\!\langle \exp_{\S}(x+y^*), P\rangle\!\rangle ,$$

and hence the complex wave representation with $P,Q \in \Gamma(\mathcal{H}+\mathcal{H}^*)$ gives the identity

$$\langle\!\langle P,Q\rangle\!\rangle = \int \overline{\tilde{P}[x+y^*]}\cdot\tilde{Q}[x+y^*]\cdot\gamma_{\mathcal{H}}^{\frac{1}{2}}(dx)\cdot\gamma_{\mathcal{H}}^{\frac{1}{2}}(dy) ,$$

where $\sim$ indicates that the functions are extended to a Hilbert–Schmidt enlargement of $\mathcal{H}+\mathcal{H}^*$ as the domain of integration.

We will examine closer the correspondence between the real wave representation of $\Gamma(\mathcal{H}+\mathcal{H}^*)$ equipped with the conjugation *, and the transformation π .

We denote by $\mathbb{H}$ the real part of $\mathcal{H}+\mathcal{H}^*$ with respect to the conjugation *, i.e.

$$\mathbb{H} = \left\{ z+z^* \mid z\in\mathcal{H} \right\} .$$

For $P \in \Gamma_0(\mathcal{H}+\mathcal{H}^*)$ we compute the ψ–value of P in $z+z^* \in \mathbb{H}$. We omit the notation $P_*(z+z^*)$ from chapter 7 and use $P(z+z^*)$ instead. We have

$$P(z+z^*) = \S P[z+z^*] = \langle\!\langle \exp_{\S}(z+z^*), \S P\rangle\!\rangle = \langle\!\langle \S:\exp(z+z^*):, \S P\rangle\!\rangle$$

$$= \langle\!\langle :\exp(z+z^*):, P\rangle\!\rangle_N = \pi P[z] .$$

The last identity we verify on the generators $af^* \in \Gamma_0(\mathcal{H}+\mathcal{H}^*)$,

$$\langle\!\langle :\exp(z+z^*):, af^* \rangle\!\rangle_N = \langle\!\langle (\exp z)\exp z^*, af^* \rangle\!\rangle_N = \langle \exp z, a\rangle \overline{\langle \exp z, f\rangle}$$
$$= a[z]\cdot\overline{f[z]} = \pi(af^*)[z] \ .$$

The real wave representation of $\Gamma(\mathcal{H}+\mathcal{H}^*)$ yields for $P,Q\in\Gamma(\mathcal{H}+\mathcal{H}^*)$

$$\langle\!\langle P,Q\rangle\!\rangle = \int \overline{\tilde{P}(z+z^*)}\cdot\tilde{Q}(z+z^*)\cdot\gamma_{\mathbb{H}}(z+z^*) \ ,$$

where the functions are extended to a Hilbert-Schmidt enlargement of $\mathbb{H}$ over which we integrate.

Let $\mathcal{H}_{\mathbb{R}}$ denote the space $\mathcal{H}$ considered as a real Hilbert space with inner product

$$\langle\cdot,\cdot\rangle_{\mathbb{R}} = \mathrm{Re}(\langle\cdot,\cdot\rangle) = \tfrac{1}{2}(\langle\cdot,\cdot\rangle + \overline{\langle\cdot,\cdot\rangle}) \ .$$

Then the unitary mapping

$$\mathcal{H}_{\mathbb{R}} \ni z \longrightarrow \frac{1}{\sqrt{2}}\,(z+z^*) \in \mathbb{H}$$

transforms the measure $\gamma_{\mathbb{H}}$ into the measure $\gamma_{\mathcal{H}}^{\frac{1}{2}}$, sitting on Hilbert-Schmidt enlargements of $\mathcal{H}$, i.e. for $P,Q \in \Gamma_0(\mathcal{H}+\mathcal{H}^*)$ we have

$$\langle\!\langle P,Q\rangle\!\rangle = \int \overline{\tilde{P}(z+z^*)}\cdot\tilde{Q}(z+z^*)\cdot\gamma_{\mathbb{H}}(z+z^*) = \int \widetilde{\pi}\overline{P[z]}\cdot\widetilde{\pi}Q[z]\cdot\gamma_{\mathcal{H}}^{\frac{1}{2}}(dz)$$

and we get the following fundamental theorem.

Theorem 9.1: The transformation π extends to a unitary mapping of $\Gamma(\mathcal{H}+\mathcal{H}^*)$ onto $L^2(\gamma_{\mathcal{H}}^{\frac{1}{2}})$.

We denote the extension by the same symbol.

The Weyl operator W_e , $e\in\mathcal{H}$, on $\Gamma_1\mathcal{H}$ is of the form

$$W_e = e^{-\frac{1}{2}\|e\|^2}\cdot a^+(\exp e)\,\exp(-e^*) \ .$$

Thus by the Campbell-Baker-Hausdorff formula and some calculations we get for $P \in \Gamma_1(\mathcal{H}+\mathcal{H}^*)$ and $z\in\mathcal{H}$

$$\pi P[z] = \langle \exp z, (\exp -z) P(\exp z)\rangle = \langle \emptyset, \exp z^* ((\exp -z) P(\exp z))\rangle$$
$$= \langle \emptyset, \exp z^* \exp -z\ P \exp z\rangle$$
$$= \langle \emptyset, W_{-z} P\, W_z \emptyset\rangle\ .$$

Thus

$$\pi P[z] = \langle \emptyset, W_{-z} P\, W_z \emptyset\rangle\ ,$$

by which πP in $z \in \mathcal{H}$ is the so-called vacuum expectation value of the operator $W_{-z} P\, W_z$.

On the generators $af^* \in \Gamma_0(\mathcal{H}+\mathcal{H}^*)$ we have

$$\pi(af^*)^*[z] = \pi(fa^*)[z] = f[z]\cdot\overline{a[z]} = \overline{a[z]\cdot\overline{f[z]}} = \overline{\pi(af^*)[z]}\ ,$$

and since the elements af^* are total in $\Gamma(\mathcal{H}+\mathcal{H}^*)$, we get

$$\pi P^* = \overline{\pi P} \quad \text{for all} \quad P \in \Gamma(\mathcal{H}+\mathcal{H}^*)\ .$$

For $P,Q \in \Gamma_1(\mathcal{H}+\mathcal{H}^*)$ and $z \in \mathcal{H}$ we have

$$\pi P[z]\cdot\pi Q[z] = P(z+z^*)\cdot Q(z+z^*) = \S P[z+z^*]\cdot \S Q[z+z^*]$$
$$= (\S P)_{\S}(\S Q)[z+z^*] = \S(:PQ:)[z+z^*]$$
$$= :PQ:(z+z^*) = \pi :PQ:[z]\ ,$$

and hence

$$\pi(:PQ:) = (\pi P)\cdot(\pi Q) \quad \text{for all} \quad P,Q \in \Gamma_1(\mathcal{H}+\mathcal{H}^*)\ .$$

We finish the chapter returning to the operators $\S$ and $\S^{-1}$ introduced in chapter 8. Consider the identity

$$\exp_{\S}(x+y^*) = \S :\exp(x+y^*): = e^{-\langle y,x\rangle} :\exp(x+y^*):$$

for $x+y^* \in \mathcal{H}+\mathcal{H}^*$.

Using the fact that $:\exp(x+y^*): = a^+(\exp x) \exp y^*$ and the Campbell–Baker–Hausdorff formula we get

$$\S^{-1} a^+(\exp x) \exp y^* = e^{\langle y,x\rangle} a^+(\exp x) \exp y^* = \exp y^*\ a^+(\exp x)\ .$$

Approximating $f,g \in \Gamma_1\mathcal{H}$ by linear combinations of coherent vectors, we show that the operator $\S^{-1}$ transforms operators in normal form, i.e. creations to the left and annihilations to the right, to the anti-normal operator, i.e.

$$\S^{-1}(fg^*) = g^* a^+(f) \quad \text{for all} \quad f,g \in \Gamma_1(\mathcal{H}+\mathcal{H}^*)\ .$$

Though it is not written in the normal form, the operator $g^{*}a^{+}(f)$ constitutes an element of $\Gamma_1(\mathcal{H}+\mathcal{H}^{*})$. To reach the normal form would require subsequent use of the canonical commutation relation.

Appendix 1: Halmos' lemma

Halmos' lemma: Let the map

$$A : \mathcal{H}_1 \longrightarrow \mathcal{H}_2$$

denote a linear contraction between two Hilbert spaces. Then there exist an orthogonal projection P, a unitary map U and an isometry J, such that

$$A = P \cdot U \cdot J$$

Proof: We explicitly define the maps.

$$J : \mathcal{H}_1 \longrightarrow \mathcal{H}_1 \oplus \mathcal{H}_2$$
$$J(x_1) = x_1 \oplus 0 .$$

$$U : \mathcal{H}_1 \oplus \mathcal{H}_2 \longrightarrow \mathcal{H}_2 \oplus \mathcal{H}_1$$
$$U = \begin{bmatrix} A & T \\ -S & A^* \end{bmatrix}$$

with $S = (I-A^*A)^{\frac{1}{2}} : \mathcal{H}_1 \longrightarrow \mathcal{H}_1$

and $T = (I-AA^*)^{\frac{1}{2}} : \mathcal{H}_2 \longrightarrow \mathcal{H}_2$.

The projection P is defined by identifying the spaces $\mathcal{H}_2$ and $\mathcal{H}_2 \oplus 0$,

$$P : \mathcal{H}_2 \oplus \mathcal{H}_1 : \longrightarrow \mathcal{H}_2$$
$$P(x_2 \oplus x_1) = x_2 .$$

It is an easy exercise to show that the maps have the required specifications. But to prove that the map U is unitary, we need to know that S and T intertwine with A

$$AS = TA .$$

This we show as follows. Consider the identity

$$A \cdot S^2 = A \cdot (I-A^*A) = (I-AA^*) \cdot A = T^2 \cdot A .$$

By a simple induction in $n\in\mathbb{N}$ this extends to

$$A\cdot S^{2n} = A\cdot S^{2(n-1)}\cdot S^2 = T^{2(n-1)}\cdot A\cdot S^2 = T^{2(n-1)}\cdot T^2\cdot A = T^{2n}\cdot A \ .$$

We have shown that

$$A\cdot \mathrm{pol}(S^2) = \mathrm{pol}(T^2)\cdot A \ ,$$

where pol denotes a polynomial.

We now choose polynomials $p_n(\cdot)$ on $[0,1]$ in such a way that $\{p_n\}_{n\in\mathbb{N}}$ converges to $\sqrt{t}$ uniformly on $[0,1]$. By the functional calculus we get

$$p_n(S^2) \xrightarrow[n]{} S \quad \text{in} \quad \mathcal{B}(\mathcal{H}_1)$$

$$p_n(T^2) \xrightarrow[n]{} T \quad \text{in} \quad \mathcal{B}(\mathcal{H}_2) \ .$$

Hence

$$AS = TA \ .$$

For $x_1\in\mathcal{H}_1$ we get

$$P\cdot U\cdot J(x_1) = P\cdot U(x_1\oplus 0) = P(Ax_1\oplus (-Sx_1)) = A(x_1) \ .$$

■

Appendix 2: Gaussian measures

The results in this appendix come from [18]. We provide the construction of the gaussian measure on infinite dimensional spaces, uniqueness of the measure and some properties that are both useful and have illustrative purposes.

The numbering is independent of the numbering in other parts of these notes.

A: The gaussian measure on $\mathbb{R}^\infty$

We define the space $\mathbb{R}^\infty$ as the set of all real sequences, i.e.

$$\mathbb{R}^\infty = \left\{ \underline{a}=(a_1,a_2,\dots,a_n,\dots) \;\middle|\; a_i\in\mathbb{R} \text{ for all } i\in\mathbb{N} \right\} .$$

We define a topology in $\mathbb{R}^\infty$ by the metric

$$d(\underline{x},\underline{y}) = |\underline{x}-\underline{y}| = \sum_{n=1}^{\infty} 2^{-n} \frac{|x_n-y_n|}{1+|x_n-y_n|} \qquad \text{where} \quad \underline{x},\underline{y}\in\mathbb{R}^\infty .$$

The well known real Hilbert space is a subset of $\mathbb{R}^\infty$,

$$\ell^2(\mathbb{R}) = \left\{ \underline{x}\in\mathbb{R}^\infty \;\middle|\; \sum_{n=1}^{\infty} x_n^2 < \infty \right\} \subset \mathbb{R}^\infty .$$

Writing $\mathbb{R}^{\infty-n}$ for the subset of $\mathbb{R}^\infty$ consisting of zeros in the first n positions, and identifying the space $\mathbb{R}^n$ as a subset of $\mathbb{R}^\infty$ by adding zeros after the first n positions, we get

$$\mathbb{R}^\infty = \mathbb{R}^n \oplus \mathbb{R}^{\infty-n} .$$

In $\mathbb{R}^\infty$ we define the cylinder sets

$$B \oplus \mathbb{R}^{\infty-n} ,$$

where n are natural numbers and B runs over Borel sets in $\mathbb{R}^n$. These cylinder sets form a base for the Borel sets of $\mathbb{R}^\infty, |\cdot|$.

We define the gaussian content γ on the cylinder sets in $\mathbb{R}^\infty$ by setting, for $n\in\mathbb{N}$ and a Borel set $B \subset \mathbb{R}^n$,

$$\gamma(B \oplus \mathbb{R}^{\infty-n}) = (2\pi)^{-n/2} \int_B \exp(-\tfrac{1}{2}(x_1^2+\dots+x_n^2))dx_1\dots dx_n$$

where $dx_1dx_2\dots dx_n$ denotes the Lebesgue integration over $\mathbb{R}^n$.

We have that

$$\gamma(\mathbb{R}^\infty) = 1 ,$$

and for every fixed $n\in\mathbb{N}$ γ becomes a measure on $\mathbb{R}^n$.

We shall use the **Kolmogorov extension theorem** saying that if γ is a content on the cylinder sets in $\mathbb{R}^\infty$, such that

$$\gamma(\mathbb{R}^\infty) = 1 ,$$

and if γ is a measure on $\mathbb{R}^n$ for every $n\in\mathbb{N}$, then γ extends uniquely to a Borel measure on $\mathbb{R}^\infty, |\cdot|$.

We denote the measure by the same symbol γ .

For $\sigma > 0$ we define

$$\gamma^\sigma(B \oplus \mathbb{R}^{\infty-n}) = (2\pi\sigma)^{-n/2} \int_B \exp(-\tfrac{1}{2}(x_1^2+..+x_n^2)/\sigma)dx_1..dx_n .$$

It is known that for different σ's the corresponding measures are pairwise singular. For simplicity we develop the theory for $\sigma = 1$.

In particular we shall need the measures for the numbers

$$\sigma = 1 \quad\text{and}\quad \sigma = \tfrac{1}{2} .$$

These measures will be denoted by γ and $\gamma^{\frac{1}{2}}$.

Theorem 1A: Consider a sequence $(a_n)_{n\in\mathbb{N}}$ of positive real numbers. We define the set

$$E = \left\{ \underline{\mathbf{x}}\in\mathbb{R}^\infty \;\middle|\; \sum_{n=1}^\infty a_n x_n^2 < \infty \right\} .$$

If the series $\sum_{n=1}^\infty a_n$ is summable then $\gamma(E)=1$, otherwise $\gamma(E)=0$.

Proof: We construct a net of functions converging to the indicator function of E , 1_E . Given $\lambda>0$, we define the function

$$f_\lambda : \mathbb{R}^\infty \longrightarrow \mathbb{R}$$

by

$$f_\lambda(\underline{\mathbf{x}}) = \exp(-\lambda \sum_{n=1}^\infty a_n x_n^2) .$$

It is easy to prove that

$$f_\lambda(\underline{x}) = \begin{cases} 0 & \text{if } \underline{x}\notin E \\ \text{a positive real number} & \text{if } \underline{x}\in E \end{cases},$$

and that $f_\lambda(\underline{x}) \longrightarrow 1_E(\underline{x})$ pointwise for $\lambda\longrightarrow 0$.

The functions

$$f_{\lambda,N}(\underline{x}) = \exp(-\lambda \sum_{n=1}^{N} a_n x_n^2) \quad \text{for } N\in\mathbb{N}$$

are obviously integrable over $\mathbb{R}^\infty$ and for fixed λ form a decreasing sequence. By the Lebesgue theorem of monotone convergence we get that

$$\int_{\mathbb{R}^\infty} f_\lambda(\underline{x})\gamma(d\underline{x}) = \int_{\mathbb{R}^\infty} \lim_{N\to\infty} f_{\lambda,N}(\underline{x})\,\gamma(d\underline{x}) = \lim_{N\to\infty} \int_{\mathbb{R}^\infty} f_{\lambda,N}(\underline{x})\gamma(d\underline{x})$$

in case the limit exists.

Notice that in the case where f depends only on a finite number of variables $x_1, x_2, \ldots, x_n$ the integration reduces to

$$\gamma(d\underline{x}) = (2\pi)^{-N/2}\exp(-\tfrac{1}{2}\sum_{n=1}^{N} x_n^2)dx_1dx_2\ldots dx_N .$$

Using the substitution $t = \sqrt{1+2\lambda a_n}\cdot x$ we compute

$$\int_{\mathbb{R}^\infty} f_{\lambda,N}(\underline{x})\gamma(d\underline{x}) = (2\pi)^{-N/2} \int_{\mathbb{R}^N} \exp(-\lambda \sum_{n=1}^{N} a_n x_n^2)\exp(-\tfrac{1}{2} \sum_{n=1}^{N} x_n^2)dx_1dx_2\ldots dx_N$$

$$= (2\pi)^{-N/2} \prod_{n=1}^{N} \left[\int_{\mathbb{R}} \exp((-\lambda a_n + \tfrac{1}{2})x_n^2)dx_n \right]$$

$$= (2\pi)^{-N/2} \prod_{n=1}^{N} \left[\int_{\mathbb{R}} \exp(-\tfrac{1}{2}t^2)\frac{dt}{\sqrt{1+2\lambda a_n}} \right]$$

$$= \prod_{n=1}^{N} (1+2\lambda a_n)^{-1/2} \qquad \text{since } \int_{\mathbb{R}} \exp(-\tfrac{1}{2}t^2)dt = \sqrt{2\pi} .$$

If $\sum_{n=1}^{\infty} a_n = \infty$, then $\prod_{n=1}^{\infty} (1+2\lambda a_n) = \infty$, thus

$$\int_{\mathbb{R}^\infty} f_\lambda(\underline{x})\gamma(d\underline{x}) = 0 .$$

Otherwise we have that

$$\sum_{n=1}^{\infty} a_n < \infty ,$$

and by the inequality for non-negative x

$$0 \le \ln(1+x) \le x$$

we get that

$$1 \le \prod_{n=1}^{\infty} (1+2\lambda a_n) < \infty .$$

This implies that

$$\int_{\mathbb{R}^\infty} f_\lambda(\underline{\mathbf{x}})\gamma(d\underline{\mathbf{x}}) = \begin{cases} 0 & \text{if} \quad \sum_{n=1}^{\infty} a_n = \infty \\ \prod_{n=1}^{\infty} (1+2\lambda a_n)^{-1/2} & \text{if} \quad \sum_{n=1}^{\infty} a_n < \infty \end{cases} .$$

By using the Lebesgue theorem of dominated convergence we get

$$\gamma(E) = \int_{\mathbb{R}^\infty} 1_E(\underline{\mathbf{x}})\gamma(d\underline{\mathbf{x}}) = \lim_{\lambda \to 0} \int_{\mathbb{R}^\infty} f_\lambda(\underline{\mathbf{x}})\gamma(d\underline{\mathbf{x}}) .$$

If $\sum_{n=1}^{\infty} a_n < \infty$, we have that

$$\lim_{\lambda \to 0} \prod_{n=1}^{\infty} (1+2\lambda a_n) = 1 ,$$

which concludes the proof.

■

The difficulty in developing the theory of gaussian measures in $\mathbb{R}^\infty$ is due to the fact that contrary to the finite dimensional case where $\gamma(\ell^2) = 1$, we have, as shown in the above, that

$$\gamma(\ell^2) = 0 .$$

To shorten the notation we define the measures γ_n on $\mathbb{R}^n$, $n \in \mathbb{N}$, by

$$\gamma_n(d\underline{\mathbf{x}}) = (2\pi)^{-\frac{1}{2}n} \cdot \exp(-\tfrac{1}{2}(x_1^2+\ldots+x_n^2))dx_1\ldots dx_n .$$

If we identify the spaces $\mathbb{R}^\infty = \mathbb{R}^n \oplus \mathbb{R}^{\infty-n}$ and $\mathbb{R}^n \times \mathbb{R}^{\infty-n}$, and denote by $\gamma_{\infty-n}$ the gaussian measure on $\mathbb{R}^{\infty-n}$, constructed in the same way as the measure γ on $\mathbb{R}^\infty$, the measure γ can be identified with the product measure $\gamma_n \otimes \gamma_{\infty-n}$,

$$\gamma = \gamma_n \otimes \gamma_{\infty-n} .$$

Definition 2A: Let f denote a function defined on $\mathbb{R}^\infty$ and integrable with respect to γ. For $n\in\mathbb{N}$ we define

$$(E_n f)(\underline{x}) = \int_{\mathbb{R}^{\infty-n}} f(\underline{x}+\underline{s})\cdot\gamma_{\infty-n}(d\underline{s}) ,$$

where $\underline{x}\in\mathbb{R}^n$ and $\underline{s}\in\mathbb{R}^{\infty-n}$, and $\mathbb{R}^n$ is treated as a subset of $\mathbb{R}^\infty$.

By the Fubini theorem $E_n f$ is well defined, and $E_n f$ is an integrable function over $(\mathbb{R}^n,\gamma_n)$ with

$$\int_{\mathbb{R}^n} (E_n f)(\underline{x})\cdot\gamma_n(d\underline{x}) = \int_{\mathbb{R}^\infty} f(\underline{s})\cdot\gamma(d\underline{s}) .$$

Lemma 3A: For $f\in L^1(\gamma)$ and $n\in\mathbb{N}$ we have that

$$\|E_n f\|_1 \le \|f\|_1 .$$

Proof: Assume that f is a non-negative function, i.e. assume that $f\ge 0$. Then we have that $E_n f \ge 0$ and

$$\|E_n f\|_1 = \int_{\mathbb{R}^n} (E_n f)(\underline{x})\cdot\gamma_n(d\underline{x}) = \int_{\mathbb{R}^\infty} f(\underline{s})\cdot\gamma(d\underline{s}) = \|f\|_1 .$$

An arbitrary f can be written uniquely as

$$f = f^+ - f^- ,$$

where f^+ and f^- are non-negative integrable functions such that

$$f^+\cdot f^- = 0$$

$$|f| = f^+ + f^- .$$

Then we have

$$\|E_n f\|_1 = \|E_n(f^+ - f^-)\|_1 \le \|E_n(f^+)\|_1 + \|E_n(f^-)\|_1 = \|f^+\|_1 + \|f^-\|_1$$
$$= \|f^+ + f^-\|_1 = \||f|\|_1 = \|f\|_1 .$$

■

Definition 4A: A function f defined on $\mathbb{R}^\infty$ is called tame if there exists an $n\in\mathbb{N}$ such that

$$f = f\circ p_n ,$$

where

$$p_n : \mathbb{R}^\infty \longrightarrow \mathbb{R}^n$$

denotes the natural projection.

Notice that if $f \in L^1(\gamma)$ is a tame function, depending only on the first n variables, then

$$E_n f = f .$$

Lemma 5A: The tame functions are dense in $L^1(\gamma)$.

Proof: Since the cylinder sets form a base for the Borel sets in $\mathbb{R}^\infty$, every Borel function can be approximated by finite linear combinations of indicator functions of cylinder sets. Indicator functions of cylinder sets are tame.

■

Theorem 6A: (Jessen) For $f \in L^1(\gamma)$ we have that

$$E_n f \xrightarrow[n \to \infty]{} f \quad \text{in} \quad L^1(\gamma) .$$

Proof: Consider $\epsilon > 0$. Find a tame function $g \in L^1(\gamma)$ such that

$$\|f - g\|_1 < \epsilon/2 .$$

As g is tame, there exists an $N \in \mathbb{N}$ such that g depends only on the first N variables. For $n \geq N$ we have that

$$E_n g = g$$

and

$$\begin{aligned} \|E_n f - f\|_1 &\leq \|E_n(f-g)\|_1 + \|f - E_n g\|_1 \\ &= \|E_n(f-g)\|_1 + \|f-g\|_1 \leq 2 \cdot \|f-g\|_1 < \epsilon . \end{aligned}$$

■

Corollary 7A: Let f denote a function defined on $\mathbb{R}^\infty$ which is integrable with respect to γ . If f is constant with respect to every finite number of variables (depending only on the tail), then f is constant a.e. in $\mathbb{R}^\infty$.

Proof: Considered as a tame function on $\mathbb{R}^\infty$, $E_n f$ depends only on the first n variables. Hence $E_n f$ is constant a.e. Since f is constant with respect to every finite number of variables, we have for all $m,n \in \mathbb{N}$ that

$$E_n f(x_1,x_2,\ldots,x_n) = \int_{\mathbb{R}^{\infty-n}} f(\underline{x}+\underline{s})\cdot\gamma_{\infty-n}(d\underline{s})$$

$$= \int_{\mathbb{R}^{\infty-m}} f(\underline{x}+\underline{s})\cdot\gamma_{\infty-m}(d\underline{s}) = E_m f(x_1,x_2,\ldots,x_m) .$$

Thus

$$f = \lim_n E_n f \quad \text{in} \quad L^1(\gamma)$$

is obviously constant a.e.

■

Remark: Since the arctan of a measurable function is integrable, corollary 7A holds for every measurable f .

The following corollary is often called the Kolmogorov zero—one law.

Corollary 8A: Let B denote a Borel set in $\mathbb{R}^\infty$. If the indicator function of B , 1_B , is constant with respect to every finite number of variables, then B is either of measure zero or one.

Proof: From corollary 7A the function 1_B is constant a.e., by which it is obvious that $1_B = 0$ a.e. or $1_B = 1$ a.e.

■

B: The linear measurable functionals on $\mathbb{R}^\infty$

Definition 1B: λ is called a linear measurable functional on $\mathbb{R}^\infty$ if λ fulfills

1) λ is defined on a linear set $E \subset \mathbb{R}^\infty$ of full measure 1 .
2) λ is linear on E and measurable with respect to γ .

In what follows we shall provide a representation for these linear measurable functionals defined on $\mathbb{R}^\infty$.

Proposition 2B: (The Kolmogorov inequality) For $n\in\mathbb{N}$ and $\underline{\mathbf{a}},\underline{\mathbf{x}}\in\mathbb{R}^\infty$ we define

$$s_n(\underline{\mathbf{x}}) = \sum_{k=1}^{n} a_k x_k .$$

For arbitrary $\epsilon>0$ we have that

$$\gamma\left(\left\{ \underline{\mathbf{x}}\in\mathbb{R}^\infty \;\middle|\; \sup_{m\leq n} \left| \sum_{k=1}^{n} a_k x_k \right| > \epsilon \right\}\right) \leq \sum_{k=1}^{n} a_k^2 / \epsilon^2 .$$

Proof: We start by noticing that

$$\frac{1}{\sqrt{2\pi}} \int_{\mathbb{R}} t\cdot\exp(-\tfrac{1}{2}t^2)dt = 0$$

and

$$\frac{1}{\sqrt{2\pi}} \int_{\mathbb{R}} t^2\exp(-\tfrac{1}{2}t^2)dt = 1 .$$

For arbitrary, pairwise disjoint, measurable sets $Q_m\subset\mathbb{R}^\infty$, $m\in\mathbb{N}$, we have

$$\sum_{k=1}^{n} a_k^2 = \sum_{k=1}^{n} a_k^2 \frac{1}{\sqrt{2\pi}} \int_{\mathbb{R}} x_k^2\cdot\exp(-\tfrac{1}{2}x_k^2)dx_k$$

$$= \sum_{k=1}^{n} \frac{1}{\sqrt{2\pi}} \int_{\mathbb{R}} a_k^2\, x_k^2\cdot\exp(-\tfrac{1}{2}x_k^2)dx_k = \sum_{k=1}^{n} \int_{\mathbb{R}^\infty} a_k^2\, x_k^2\, \gamma(d\underline{\mathbf{x}})$$

$$= \int_{\mathbb{R}^\infty} \Big(\sum_{k=1}^{n} a_k x_k\Big)^2 \gamma(d\underline{\mathbf{x}}) = \int_{\mathbb{R}^\infty} (s_n(\underline{\mathbf{x}}))^2 \gamma(d\underline{\mathbf{x}}) ,$$

and since the relation $s_n^2 = s_m^2 + 2s_m(s_n-s_m) + (s_n-s_m)^2$ implies the estimate $s_n^2 \geq s_m^2 + 2s_m(s_n-s_m)$ we can continue

$$\geq \sum_{m=1}^{n} \int_{Q_m} [s_m^2 + 2s_m(s_n-s_m)]\gamma(d\underline{\mathbf{x}}) .$$

We construct the sets Q_m, $m\in\mathbb{N}$, as the inverse images by the function

$$f(\underline{\mathbf{x}}) = \text{the smallest } k\in\mathbb{N} \text{ such that } |s_k(\underline{\mathbf{x}})| > \epsilon$$
$$= \inf \left\{ k\in\mathbb{N} \;\middle|\; |s_k(\underline{\mathbf{x}})| > \epsilon \right\} .$$

Hence

$$Q_m = f^{-1}(m) = \left\{ \underline{\mathbf{x}}\in\mathbb{R}^\infty \;\middle|\; f(\underline{\mathbf{x}}) = m \right\} .$$

Notice that the function 1_{Q_m} depends only on the first m coordinates $x_1, x_2, .., x_m$.

We return to the integration. Since the functions s_m and 1_{Q_m} depend only on the first m variables $x_1, x_2, ..., x_m$, and the function $(s_n - s_m)$ depends only on the variables $x_{m+1}, x_{m+2}, ..., x_n$, we get by the Fubini theorem

$$\int_{Q_m} s_m(s_n - s_m)\gamma(d\underline{\mathbf{x}}) = \int_{\mathbb{R}^\infty} (1_{Q_m} \cdot s_m)(s_n - s_m)\gamma(d\underline{\mathbf{x}})$$
$$= \int_{\mathbb{R}^\infty} 1_{Q_m} \cdot s_m \gamma(d\underline{\mathbf{x}}) \int_{\mathbb{R}^\infty} (s_n - s_m)\gamma(d\underline{\mathbf{x}}) = 0 ,$$

$s_n - s_m$ being linear in the variables $x_{m+1}, x_{m+2}, .., x_n$.

We estimate the second expression using the fact that if $\underline{\mathbf{x}}\in Q_m$, then $f(\underline{\mathbf{x}}) = m$, and hence $s_m(\underline{\mathbf{x}}) > \epsilon$. Then we get

$$\int_{Q_m} s_m^2 \gamma(d\underline{\mathbf{x}}) \geq \int_{Q_m} \epsilon^2 \gamma(d\underline{\mathbf{x}}) = \epsilon^2 \gamma(Q_m) .$$

We return to the first estimate

$$\sum_{k=1}^{n} a_k^2 \geq \sum_{m=1}^{n} \int_{Q_m} s_m^2 \gamma(d\underline{\mathbf{x}}) \geq \sum_{m=1}^{n} \epsilon^2 \gamma(Q_m) = \epsilon^2 \gamma(\bigcup_{m=1}^{n} Q_m)$$
$$= \epsilon^2 \gamma(\left\{ \underline{\mathbf{x}}\in\mathbb{R}^\infty \;\middle|\; \text{there exist } m\leq n \text{ such that } \underline{\mathbf{x}}\in Q_m \right\})$$
$$= \epsilon^2 \gamma(\left\{ \underline{\mathbf{x}}\in\mathbb{R}^\infty \;\middle|\; \text{there exist } m\leq n \text{ such that } f(\underline{\mathbf{x}})=m \right\})$$
$$= \epsilon^2 \gamma(\left\{ \underline{\mathbf{x}}\in\mathbb{R}^\infty \;\middle|\; \sup_{m\leq n} |s_m(\underline{\mathbf{x}})| > \epsilon \right\}) ,$$

which concludes the proof.

■

Theorem 3B: (The Kolmogorov large number theorem) Consider a sequence of positive numbers $(a_n)_{n\in\mathbb{N}}$. If $\sum_{n=1}^{\infty} a_n^2 < \infty$, then the series

$$\lambda(\underline{\mathbf{x}}) = \sum_{n=1}^{\infty} a_n x_n$$

converges a.e. in $\mathbb{R}^{\infty}$ with respect to γ .

Proof: We shall prove that the measure of

$$\left\{ \underline{\mathbf{x}} \in \mathbb{R}^{\infty} \;\middle|\; \sum_{n=1}^{\infty} a_n x_n \text{ diverges} \right\}$$

is zero.

We define
$$s_n(\underline{\mathbf{x}}) = \sum_{k=1}^{n} a_k x_k$$
$$\bar{s}(\underline{\mathbf{x}}) = \limsup_n s_n(\underline{\mathbf{x}})$$
$$\underline{s}(\underline{\mathbf{x}}) = \liminf_n s_n(\underline{\mathbf{x}}) .$$

For $m \in \mathbb{N}$ we have

$$\begin{aligned}
\bar{s}(\underline{\mathbf{x}}) - \underline{s}(\underline{\mathbf{x}}) &= \limsup_n s_n(\underline{\mathbf{x}}) - \liminf_n s_n(\underline{\mathbf{x}}) \\
&\leq \sup_{p\geq m} s_p(\underline{\mathbf{x}}) - \inf_{q\geq m} s_q(\underline{\mathbf{x}}) = \sup_{p\geq m} s_p(\underline{\mathbf{x}}) + \sup_{q\geq m} -s_q(\underline{\mathbf{x}}) \\
&= \sup_{p,q\geq m} \left\{ s_p(\underline{\mathbf{x}}) - s_q(\underline{\mathbf{x}}) \right\} \leq \sup_{p,q\geq m} |s_p(\underline{\mathbf{x}}) - s_q(\underline{\mathbf{x}})| \\
&\leq \sup_{p\geq m} |s_p(\underline{\mathbf{x}}) - s_m(\underline{\mathbf{x}})| + \sup_{q\geq m} |s_m(\underline{\mathbf{x}}) - s_q(\underline{\mathbf{x}})| \\
&= 2 \sup_{p\geq m} |s_p(\underline{\mathbf{x}}) - s_m(\underline{\mathbf{x}})| .
\end{aligned}$$

For arbitrary $\epsilon > 0$ we define

$$M = \left\{ \underline{\mathbf{x}} \in \mathbb{R}^{\infty} \;\middle|\; |\bar{s}(\underline{\mathbf{x}}) - \underline{s}(\underline{\mathbf{x}})| > 2\epsilon \right\} .$$

Hence for all $m \in \mathbb{N}$

$$M \subset M_m := \left\{ \underline{\mathbf{x}} \in \mathbb{R}^{\infty} \;\middle|\; \sup_{p\geq m} |\bar{s}(\underline{\mathbf{x}}) - \underline{s}(\underline{\mathbf{x}})| > \epsilon \right\}$$

and thus

$$\gamma(M) \leq \gamma(M_m) \quad \text{for all } m \in \mathbb{N} .$$

Using the Kolmogorov inequality, we compute

$$\gamma(M) \le \limsup_m \gamma(M_m)$$
$$= \limsup_m \gamma\left(\left\{ \underline{x}\in\mathbb{R}^\infty \;\middle|\; \sup_{n\ge m} |s_n(\underline{x}) - s_m(\underline{x})| > \epsilon \right\}\right)$$
$$\le \limsup_m \sum_{k=m}^{\infty} a_k^2 /\epsilon^2 = 0 .$$

■

We have seen that series of the form

$$\lambda(\underline{x}) = \sum_{n=1}^{\infty} a_n x_n \quad \text{with} \quad \sum_{n=1}^{\infty} a_n^2 < \infty$$

converges a.e. in $\mathbb{R}^\infty$, and thus defines a linear measurable functional. The converse is true as well.

We introduce some notation. Let us define

$$e_n = (0, .. ,0,1,0, ..) ,$$

where the number 1 takes the n–th position , and

$$<\underline{x},e_n> = x_n \quad \text{for} \quad \underline{x} = (x_1,x_2,..,x_n,..)\in\mathbb{R}^\infty .$$

Theorem 4B: Consider a linear measurable functional λ defined on $\mathbb{R}^\infty$. Then we have

1) $\sum_{n=1}^{\infty} |\lambda(e_n)|^2 < \infty$.

2) $\lambda(\underline{x}) = \sum_{n=1}^{\infty} \lambda(e_n)<\underline{x},e_n>$ a.e. in $\mathbb{R}^\infty$.

Moreover, the representation 2) is unique.

Before proving the theorem we need some preparations. We define the translation operator

$$T_{\underline{y}} \quad \text{for } \underline{y}\in\mathbb{R}^\infty$$

by setting

$$(T_{\underline{y}}f)(\underline{x}) = f(\underline{x}-\underline{y}) ,$$

where f denotes a function defined on $\mathbb{R}^\infty$. Moreover, we introduce a subspace of $\mathbb{R}^{\infty}$,

$$\mathbb{R}_0^{\infty} = \left\{ \underline{x} \in \mathbb{R}^{\infty} \;\middle|\; \text{there exists an } N \in \mathbb{N} \text{ such that } x_n = 0 \text{ for } n \geq N \right\} .$$

Theorem 5B: Consider a measurable set $B \subset \mathbb{R}^{\infty}$. Then the following statements are equivalent,

(1) $$\gamma(B) > 0 .$$

(2) $$\gamma(\underline{y} + B) > 0 \text{ for all } \underline{y} \in \mathbb{R}_0^{\infty} .$$

Proof: It is easy to see that the theorem holds in the case where γ is the gaussian measure γ_n in the finite dimensional space $\mathbb{R}^n$.

Let us first prove the restricted form of theorem 5B, where B are cylinder sets. Consider the set

$$C = B \oplus \mathbb{R}^{\infty - n}$$

for an $n \in \mathbb{N}$ and a Borel set $B \subset \mathbb{R}^n$. Assume that $\gamma(C) > 0$. For arbitrary $\underline{y} \in \mathbb{R}_0^{\infty}$ there exists an $m \in \mathbb{N}$ such that $\underline{y} \in \mathbb{R}^m$. We have to consider two cases.

$m \leq n$: Then we regard $\underline{y}$ as being in the space $\mathbb{R}^n$, and the result follows by the finite dimensional case.

$m > n$: We rewrite the set C as

$$C = B \oplus \mathbb{R}^{\infty - n} = B \oplus \mathbb{R}^{m-n} \oplus \mathbb{R}^{\infty - m} ,$$

where $B \oplus \mathbb{R}^{m-n}$ is a Borel set in $\mathbb{R}^m$. Then we refer to the case $m \leq n$.

To prove the full version of the theorem, take a non-negative integrable function $f : \mathbb{R}^{\infty} \longrightarrow \mathbb{R}$. Consider $\underline{y} \in \mathbb{R}_0^{\infty}$, then there exists an $m \in \mathbb{N}$ such that $\underline{y} \in \mathbb{R}^m$. If $n \geq m$ we have

$$T_{\underline{y}} E_n(f) = E_n T_{\underline{y}}(f) ,$$

where $$(E_n f)(z_1, z_2, \ldots, z_n) = \int_{\mathbb{R}^{\infty - n}} f(\underline{x} + \underline{z}) \gamma_{\infty - n}(d\underline{x})$$

and $$\underline{z} = (z_1, z_2, \ldots, z_n) .$$

By the Fubini theorem the function

$$E_n f : \mathbb{R}^n \longrightarrow \mathbb{C}$$

is integrable over $\mathbb{R}^n$ with respect to γ_n and

$$\int_{\mathbb{R}^n} (E_n f)(\underline{z})\gamma_n(d\underline{z}) = \int_{\mathbb{R}^\infty} f(\underline{x})\gamma(d\underline{x}) .$$

Notice that $f \geq 0$ implies that $E_n f \geq 0$. Assume that

$$\int_{\mathbb{R}^\infty} f(\underline{x})\gamma(d\underline{x}) > 0 .$$

Hence

$$\int_{\mathbb{R}^n} (E_n f)(\underline{z})\gamma(d\underline{z}) > 0 \quad \text{for all} \quad n \in \mathbb{N} .$$

Using the theorem for the finite dimensional case and the relation

$$T_{\underline{Y}} E_n(f) = E_n T_{\underline{Y}}(f) ,$$

we get that

$$\int_{\mathbb{R}^n} (E_n T_{\underline{Y}} f)(\underline{z})\gamma(d\underline{z}) > 0 \quad \text{for all} \quad n \in \mathbb{N} .$$

Hence

$$0 < \int_{\mathbb{R}^n} (E_n T_{\underline{Y}} f)(\underline{z})\gamma(d\underline{z}) = \int_{\mathbb{R}^\infty} T_{\underline{Y}} f(\underline{x})\gamma(d\underline{x}) .$$

Setting $f = 1_B$, the theorem follows.

■

Corollary 6B: For arbitrary linear measurable functional λ we have that

$$\mathbb{R}_0^\infty \subset \mathfrak{D}(\lambda) ,$$

where $\mathfrak{D}(\lambda)$ denotes the domain of λ.

Proof: $\mathfrak{D}(\lambda)$ contains a linear subset E of full measure. We will prove that

$$\mathbb{R}_0^\infty \subset E \subset \mathfrak{D}(\lambda) .$$

Assume that there exists an $\underline{x}_0 \in \mathbb{R}_0^\infty \setminus E$. For positive t we define the sets

$$E_t = t\underline{x}_0 + E = \{t\underline{x}_0 + \underline{x} \mid \underline{x} \in E\} .$$

Since E is a linear set, the sets E_t are pairwise disjoint for different indices. Using theorem 5B, we get

$$\gamma(E_t) > 0 .$$

Hence E_t is a family of pairwise disjoint sets with positive measures. This contradicts the fact that $\gamma(\mathbb{R}^\infty) = 1$.

■

Proposition 7B: Denote by λ a linear measurable functional on $\mathbb{R}^\infty$. Then we have that

$$\sum_{n=1}^{\infty} |\lambda(e_n)|^2 < \infty .$$

Proof: For $\underline{\mathbf{x}} = (x_1,x_2,..,x_n,..)\in\mathbb{R}^\infty$ we define

$$\underline{\mathbf{x}}_{(n)} = (0,0,..,0,x_{n+1},x_{n+2},..)\in\mathbb{R}^\infty .$$

Thus we have that

$$\underline{\mathbf{x}} = \sum_{k=1}^{n} x_k e_k + \underline{\mathbf{x}}_{(n)} .$$

For $\underline{\mathbf{x}}\in\mathfrak{D}(\lambda)$ we get that

$$\lambda(\underline{\mathbf{x}}) = \sum_{k=1}^{n} x_k\lambda(e_k) + \lambda(\underline{\mathbf{x}}_{(n)}) = \sum_{k=1}^{n} x_k\lambda(e_k) + \lambda_n(\underline{\mathbf{x}}) ,$$

where $\lambda_n(\underline{\mathbf{x}}) = \lambda(\underline{\mathbf{x}}_{(n)})$ is measurable for all n , and has the same domain of definition as λ . Hence $\exp(i\cdot\lambda(\underline{\mathbf{x}}))$ and $\exp(i\cdot\lambda_n(\underline{\mathbf{x}}))$ are integrable over $\mathbb{R}^\infty$.

For arbitrary $u > 0$ we get

$$\int_{\mathbb{R}^\infty} \exp(i\cdot u\cdot\lambda(\underline{\mathbf{x}}))\gamma(d\underline{\mathbf{x}}) = \int_{\mathbb{R}^\infty} \exp\Big[i\cdot u \sum_{k=1}^{n} \lambda(e_k)x_k\Big]\exp(i\cdot u\cdot\lambda_n(\underline{\mathbf{x}}))\gamma(d\underline{\mathbf{x}})$$

$$= \prod_{k=1}^{n} \int_{\mathbb{R}} \exp\Big[i\cdot u\cdot\lambda(e_k)t - \tfrac{1}{2}t^2\Big]\frac{dt}{\sqrt{2\pi}} \cdot \int_{\mathbb{R}^\infty} \exp(i\cdot u\cdot\lambda_n(\underline{\mathbf{x}}))\gamma(d\underline{\mathbf{x}})$$

$$= \prod_{k=1}^{n} \exp(-\tfrac{1}{2}u^2|\lambda(e_k)|^2) \cdot \int_{\mathbb{R}^\infty} \exp(i\cdot u\cdot\lambda_n(\underline{\mathbf{x}}))\gamma(d\underline{\mathbf{x}}) .$$

Elementary computation ascertains that

$$\exp(-\tfrac{1}{2}u^2|\lambda(e_k)|^2) = \int_{\mathbb{R}} \exp\left[i\cdot u\cdot\lambda(e_k)t - \tfrac{1}{2}t^2\right]\frac{dt}{\sqrt{2\pi}} .$$

Using this we get that

$$\left|\int_{\mathbb{R}^\infty} \exp(i\cdot u\cdot\lambda(\underline{\mathbf{x}}))\gamma(d\underline{\mathbf{x}})\right| \leq \exp(-\tfrac{1}{2}u^2 \sum_{k=1}^{n} |\lambda(e_k)|^2) \quad \text{for all} \quad n .$$

Assume that $\sum_{n=1}^{\infty} |\lambda(e_n)|^2 = \infty$. This yields

$$\int_{\mathbb{R}^\infty} \exp(i\cdot u\cdot\lambda(\underline{\mathbf{x}}))\gamma(d\underline{\mathbf{x}}) = 0$$

for all $u > 0$. By the dominated convergence theorem we get

$$\lim_{n\to\infty} \int_{\mathbb{R}^\infty} \exp(i\cdot\tfrac{1}{n}\lambda(\underline{\mathbf{x}}))\gamma(d\underline{\mathbf{x}}) = \int_{\mathbb{R}^\infty} \lim_{n\to\infty} \exp(i\cdot\tfrac{1}{n}\lambda(\underline{\mathbf{x}}))\gamma(d\underline{\mathbf{x}})$$

$$= \int_{\mathbb{R}^\infty} 1\cdot\gamma(d\underline{\mathbf{x}}) = 1 ,$$

which is a contradiction.

■

Proposition 8B: Consider a linear measurable functional λ . If

$$\lambda(e_n) = 0 \quad \text{for all} \quad n\in\mathbb{N} ,$$

then

$$\lambda = 0 \quad \text{a.e.}$$

Proof: $\lambda(e_n) = 0$ for all $n\in\mathbb{N}$ implies that

$$\lambda|_{\mathbb{R}_0^\infty} = 0 .$$

Let E denote the domain of definition for λ , and define

$$E^+ = \{ \underline{\mathbf{x}}\in E \mid \lambda(\underline{\mathbf{x}}) \geq 0 \}$$

$$E^- = \{ \underline{\mathbf{x}}\in E \mid \lambda(\underline{\mathbf{x}}) \leq 0 \} .$$

Since λ is linear, we have that $E^+ = -E^-$, and since the measure γ is symmetric, we get

$$\gamma(E^+) = \gamma(E^-) .$$

Then it is obvious that

$$\gamma(E^+) + \gamma(E^-) \geq \gamma(E) = 1 .$$

Since $\underline{\mathbf{x}} + E^+ = E^+$ for every $\underline{\mathbf{x}}\in\mathbb{R}_0^\infty$ and likewise for E^- , both 1_{E^+} and 1_{E^-} are constant with respect to every finite number of variables. From the Kolmogorov zero-one law we get that $\gamma(E^+)$ is either 0 or 1, and likewise with E^-. Then

$$\gamma(E^+) = \gamma(E^-) = 1 ,$$

and hence

$$\gamma(\{ \underline{\mathbf{x}} \in E \mid \lambda(\underline{\mathbf{x}}) = 0 \}) = \gamma(E^+ \cap E^-) = 1 .$$

■

We are now ready to prove theorem 4B.

<u>**Proof (Theorem 4B):**</u> Since 1) amounts to proposition 7B, it remains to prove 2), i.e.

$$\lambda(\underline{\mathbf{x}}) = \sum_{n=1}^{\infty} \lambda(e_n)\langle\underline{\mathbf{x}},e_n\rangle \quad \text{a.e.}$$

Since by proposition 7B

$$\sum_{n=1}^{\infty} |\lambda(e_n)|^2 < \infty ,$$

theorem 3B ascertains that

$$\sum_{n=1}^{\infty} \lambda(e_n)\langle\underline{\mathbf{x}},e_n\rangle$$

converges a.e. in $\mathbb{R}^\infty$ and defines a linear measurable functional Λ. We must prove that $\Lambda = \lambda$. For $k \in \mathbb{N}$ we have

$$\Lambda(e_k) = \sum_{n=1}^{\infty} \lambda(e_n)\langle e_k,e_n\rangle = \lambda(e_k) .$$

By proposition 8B we conclude

$$\lambda(\underline{\mathbf{x}}) = \Lambda(\underline{\mathbf{x}}) = \sum_{n=1}^{\infty} \lambda(e_n)\langle\underline{\mathbf{x}},e_n\rangle \quad \text{a.e.}$$

■

For arbitrary functionals $\lambda_1,\ldots,\lambda_n$, we wish to calculate the measure of sets of the form

$$\left\{ \underline{\mathbf{x}} \in \mathbb{R}^\infty \;\middle|\; a_k < \lambda_k(\underline{\mathbf{x}}) \leq b_k \ , \ k=1,..,n \right\} ,$$

where a_k, b_k are real numbers. We shall often use the shorter notation

$$[\lambda > c] = \left\{ \underline{\mathbf{x}} \in \mathbb{R}^\infty \;\middle|\; \lambda(\underline{\mathbf{x}}) > c \right\} .$$

Lemma 9B: Consider a sequence $\{f_n\}_{n=1}^\infty$ of measurable functions in $\mathbb{R}^\infty$ with values in $\mathbb{R}$. If $\{f_n\}_{n=1}^\infty$ converges almost everywhere to a function f, i.e.

$$f_n \xrightarrow[n]{} f \quad \text{pointwise for a.e. } \underline{\mathbf{x}} \in \mathbb{R}^\infty ,$$

then to every $c \in \mathbb{R}$ there exists a sequence $\{c_k\}_{k \in \mathbb{N}}$ fulfilling

$$c_k \xrightarrow[k]{} c$$

and

$$\gamma[f > c] = \lim_k \lim_n \gamma[f_n \geq c_k] .$$

Proof: Assume that $[f=c] = 0$ and denote by M the zero-measure set on which $\{f_n\}_{n=1}^\infty$ does not converge to f. Then we get

$$1_{[f_n \geq c]} \xrightarrow[n]{} 1_{[f>c]}$$

pointwise on the set $\mathbb{R}^\infty \setminus (M \cup [f=c])$, i.e. almost everywhere on $\mathbb{R}^\infty$.

By using the dominated convergence theorem we get

$$\gamma[f_n \geq c] \xrightarrow[n]{} \gamma[f > c] .$$

The case when the set $[f=c]$ has positive measure now follows. We find a sequence $\{c_k\}_{k \in \mathbb{N}}$ of real numbers fulfilling

$$c_k \xrightarrow[k]{} c \quad \text{and} \quad \gamma[f = c_k] = 0 \quad \text{for every } k \in \mathbb{N} .$$

This is possible, since there would otherwise exist an uncountable family of disjoint sets with positive measure, contradicting the fact that the measure γ is finite.

By applying the above established result to the zero–measure sets $[f>c_k]$ and using the dominated convergence theorem, we get,

$$\gamma[f>c] = \lim_{k\to\infty} \gamma[f>c_k] = \lim_{k\to\infty} \lim_{n\to\infty} \gamma[f_n\geq c_k] .$$

∎

The measure of the sets

$$\gamma[f_1>c_1,\ f_2>c_2\ ,\dots,\ f_m>c_m] ,$$

where $f_1,f_2,\dots,f_m$ are measurable functions, can be calculated in a similar way.

We shall apply the lemma with a linear measurable functional in $\mathbb{R}^\infty$, denoted by λ. It has been proved earlier that

$$\lambda(\underline{\mathbf{x}}) = \sum_{n=1}^{\infty} a_n x_n \quad \text{with} \quad \sum_{n=1}^{\infty} a_n^2 < \infty .$$

We denote by λ_N the sum of the terms with indices from 1 to N. Then by lemma 9B we get

$$\gamma[\lambda>c] = \lim_{k\to\infty} \lim_{N\to\infty} \gamma[\lambda_N\geq c_k] ,$$

where the sequence $\{c_k\}_{k\in\mathbb{N}}$ converges to $c\in\mathbb{R}$.

We now calculate $\gamma[\lambda_N\geq c_k]$.

$$\gamma[\lambda_n\geq c_m] = \gamma\left\{ \underline{\mathbf{x}}\in\mathbb{R}^\infty \;\middle|\; \sum_{k=1}^{n} a_k x_k \geq c_m \right\}$$

$$= (2\pi)^{-\frac{1}{2}n} \int_{\mathbb{R}^n} 1_M\cdot\exp(-\tfrac{1}{2}(x_1^2+\dots+x_n^2))dx_1\dots dx_n ,$$

where

$$M = \left\{ \underline{\mathbf{x}}\in\mathbb{R}^n \;\middle|\; \sum_{k=1}^{n} a_k x_k \geq c_m \right\} .$$

By choosing an orthogonal transformation in $\mathbb{R}^n$ sending the line spanned by

$$(\ a_1,a_2,\dots,a_n\ )/\|\underline{\mathbf{a}}_n\|^2 ,$$

$\|\underline{\mathbf{a}}_n\|^2 = \sum_{k=1}^{n} a_k^2$, into the line spanned by the first natural basis vector in $\mathbb{R}^n$, we get by using the transformation theorem

$$\gamma[\lambda_n \geq c_m] = (2\pi)^{-\frac{1}{2}} \int\limits_{\{t \geq c_m/\|\underline{a}_n\|^2\}} \exp(-\tfrac{1}{2}t^2)dt = (2\pi)^{-\frac{1}{2}} \int_{c_m/\|\underline{a}_n\|^2}^{\infty} \exp(-\tfrac{1}{2}t^2)dt .$$

We choose λ with $\|\lambda\|_2^2 = 1$, i.e. $\sum_{n=1}^{\infty} a_n^2 = 1$. By first letting n , and afterwards m , go to infinity the above expression reduces to

$$\gamma[\lambda > c] = (2\pi)^{-\frac{1}{2}} \int_{c}^{\infty} \exp(-\tfrac{1}{2}t^2)dt ,$$

with λ being a linear measurable functional in $\mathbb{R}^{\infty}$ with

$$\sum_{n=1}^{\infty} |\lambda(e_n)|^2 = 1 .$$

The expression can easily be extended to

$$\gamma[\lambda_1 > c_1, .., \lambda_m > c_m] = (2\pi)^{-\frac{1}{2}m} \int_{c_1}^{\infty} \dots \int_{c_m}^{\infty} \exp(-\tfrac{1}{2}(x_1^2 + .. + x_m^2))dx_1 .. dx_m ,$$

where $\lambda_1, .., \lambda_m$ denote linear measurable functionals in $\mathbb{R}^{\infty}$, all with

$$\sum_{n=1}^{\infty} |\lambda_i(e_n)|^2 = 1 \quad \text{for} \quad i=1,..,m$$

and the vectors $\{a_n^i = \lambda_i(e_n)\}_{n \in \mathbb{N}}$ orthogonal for different indices i . One hereby obtains an orthogonal transformation in $\mathbb{R}^N$ with $N \geq m$.

A simple set theoretical argument together with the additive property of the measure give us the expression

$$\gamma[a_1 < \lambda_1 \leq b_1, .., a_m < \lambda_m \leq b_m] = (2\pi)^{-\frac{1}{2}m} \int_{a_1}^{b_1} \dots \int_{a_m}^{b_m} \exp(-\tfrac{1}{2}(x_1^2 + .. + x_m^2))dx_1 .. dx_m ,$$

where $\lambda_1, .., \lambda_m$ are measurable functionals in $\mathbb{R}^{\infty}$ with

$$\sum_{n=1}^{\infty} |\lambda_i(e_n)|^2 = 1 \quad \text{for} \quad i=1,..,m ,$$

and $\{a_n^i = \lambda_i(e_n)\}_{n \in \mathbb{N}}$ orthogonal for different indices i .

C: Linear transformations in $\mathbb{R}^{\infty}$

We shall extend unitary transformations of ℓ^2 onto ℓ^2 to maps in $\mathbb{R}^{\infty}$ that preserve the gaussian measure.

Definition 1C: A weak measurable linear transformation in $\mathbb{R}^{\infty}$ is a map

$$A : \mathbb{R}^{\infty} \supseteq E_A \ni \underline{x} \longrightarrow \underline{y} = (y_1, y_2, \ldots, y_m, \ldots) = A\underline{x} \in \mathbb{R}^{\infty} ,$$

where

1) The domain of definition E_A is a linear set of full measure.

2) The map A is linear.

3) Every coordinate function $y_m : E_A \longrightarrow \mathbb{R}$ is a linear measurable functional in $\mathbb{R}^{\infty}$.

There are some comments in connection with this definition.

1) According to the earlier paragraph concerning functionals in $\mathbb{R}^{\infty}$ we have that

$$y_m(\underline{x}) = \sum_{n=1}^{\infty} a_{mn} x_n \quad \text{with} \quad \{a_{mn}\}_{n \in \mathbb{N}} \in \ell^2 .$$

The transformation A is then given by the (infinite) matrix $\{a_{mn}\}_{m,n \in \mathbb{N}}$ where the rows $\{a_{mn}\}_{n \in \mathbb{N}}$ are elements in ℓ^2 .

2) By the maximal domain for a linear functional λ in $\mathbb{R}^{\infty}$ we understand the set

$$\left\{ \underline{x} \in \mathbb{R}^{\infty} \;\middle|\; \sum_{n=1}^{\infty} \lambda(e_n) x_n \text{ converges} \right\} .$$

If we let E_m denote the maximal domain for the functional y_m and set $E = \bigcap_{m=1}^{\infty} E_m$, we get that

$$E_A \subset E \quad \text{and} \quad \gamma(E) = 1 .$$

If $E_A \neq E$, we extend in the obvious manner the transformation A to the whole E , thus getting A maximally defined.

3) It follows from the earlier results that $\ell^2 \subset E_A$ and that A is uniquely determined by its values on ℓ^2. It is indeed determined by its values on the set

$$\{e_n\}_{n\in\mathbb{N}},$$

where

$$e_n = (0,0,..,0,1,0...) .$$

$$\uparrow$$

coordinate number n

4) The converse holds as well; if elements of ℓ^2 are taken as the rows of a matrix $A = \{a_{mn}\}_{m,n\in\mathbb{N}}$, then A is a weak measurable linear transformation in $\mathbb{R}^\infty$.

Let A denote a linear bounded transformation

$$A : \ell^2 \longrightarrow \ell^2 .$$

Then we extend A uniquely to a weak measurable linear transformation in $\mathbb{R}^\infty$ by extending the domain of definition for the matrix $\{a_{mn}\}_{m,n\in\mathbb{N}}$ given by

$$a_{mn} = \langle Ae_n, e_m\rangle .$$

Since the rows are elements of ℓ^2, by the Parseval identity

$$\sum_{n=1}^{\infty} a_{mn}^2 = \sum_{n=1}^{\infty} |\langle Ae_n, e_m\rangle|^2 = \sum_{n=1}^{\infty} |\langle e_n, A^* e_m\rangle|^2 = \|A^* e_m\|^2 < \infty$$

the extension is well defined.

Theorem 2C: Let $U : \ell^2 \longrightarrow \ell^2$ denote a unitary transformation. We extend U to a weak measurable linear map in $\mathbb{R}^\infty$ by the matrix $\{u_{mn} = \langle Ue_n, e_m\rangle\}_{m,n\in\mathbb{N}}$. Then the gaussian measure γ is invariant under the transformation U.

Proof: Consider $n\in\mathbb{N}$ and $a_k, b_k \in \mathbb{R}$, $k=1,..,n$. We define

$$B = \left\{ \underline{x}\in\mathbb{R}^\infty \;\middle|\; a_k < x_k \le b_k \text{ for } k=1,..,n \right\} .$$

We have to show that B and U(B) are sets of equal measure.

It is obvious that

$$\gamma(B) = (2\pi)^{-\frac{1}{2}n} \int_{a_1}^{b_1} \cdots \int_{a_n}^{b_n} \exp(-\tfrac{1}{2}(x_1^2+\cdots+x_n^2))\,dx_1\cdots dx_n$$

and that

$$U(B) = \left\{ \underline{\mathbf{x}} \in \mathbb{R}^\infty \;\middle|\; a_k < (U^*\underline{\mathbf{x}})_k \le b_k \quad \text{for } k=1,..,n \right\},$$

where U^* denote the extension of the adjoint of U. Since the rows in the matrix of U are the columns in the matrix of U^*, we get that

$$U(B) = \left\{ \underline{\mathbf{x}} \in \mathbb{R}^\infty \;\middle|\; a_k < \sum_{i=1}^{\infty} u_{ik}x_i \le b_k \quad \text{for } k=1,..,n \right\}.$$

We define the functionals

$$\lambda_k(\underline{\mathbf{x}}) = \sum_{i=1}^{\infty} u_{ik}x_i \qquad k=1,..,n\,,$$

and observe that since the rows in U are pairwise orthogonal,

$$\begin{aligned}\gamma(U(B)) &= \gamma\left\{ \underline{\mathbf{x}} \in \mathbb{R}^\infty \;\middle|\; a_k < \lambda_k(\underline{\mathbf{x}}) \le b_k \quad \text{for } k=1,..,n \right\} \\ &= (2\pi)^{-\frac{1}{2}n} \int_{a_1}^{b_1} \cdots \int_{a_n}^{b_n} \exp(-\tfrac{1}{2}(x_1^2+\cdots+x_n^2))\,dx_1\cdots dx_n\,.\end{aligned}$$

■

D: The gaussian measure on $\mathbb{C}^\infty$

Consider the linear space

$$\mathbb{C}^\infty = \left\{ \underline{\mathbf{z}} = (z_1, z_2, \ldots, z_n, \ldots) \;\middle|\; z_i \in \mathbb{C} \quad \text{for all} \quad i \in \mathbb{N} \right\}$$

and the Hilbert space

$$\ell^2(\mathbb{C}) = \left\{ \underline{\mathbf{z}} \in \mathbb{C}^\infty \;\middle|\; \sum_{n=1}^{\infty} |z_n|^2 < \infty \right\},$$

and introduce in $\mathbb{C}^n$, $n \in \mathbb{N}$, the gaussian measure

$$\int_{\mathbb{C}^n} f(\underline{\mathbf{z}})\cdot\gamma^{\frac{1}{2}}(d\underline{\mathbf{z}}) = \pi^{-n}\int_{\mathbb{R}^{2n}} f(\underline{\mathbf{z}})\cdot e^{-|\underline{\mathbf{z}}|^2}\,dx_1dy_1\ldots dx_ndy_n$$

where

$\underline{z} = (z_1, z_2, \ldots, z_n) \in \mathbb{C}^n$ and $z_k = x_k + i\cdot y_k \in \mathbb{C}$

$|\underline{z}|^2 = |z_1|^2 + \ldots + |z_n|^2$

and $dx_1 dy_1 \ldots dx_n dy_n$ indicates the Lebesgue integration in $\mathbb{R}^{2n}$.

Then all former results concerning the gaussian measure are easily extended to this new gaussian measure on $\mathbb{C}^\infty$.

E: Hilbert–Schmidt enlargements

Let $\mathcal{H}, \langle , \rangle$ denote a real or complex Hilbert space. A Hilbert–Schmidt enlargement $\tilde{\mathcal{H}}, \tilde{\langle , \rangle}$ of $\mathcal{H}, \langle , \rangle$ is itself a new Hilbert space containing $\mathcal{H}$ as a linear dense subset and such that the inclusion mapping of $\mathcal{H}$ into $\tilde{\mathcal{H}}$ is a Hilbert–Schmidt operator, i.e.

$$\sum_{n=1}^{\infty} {}^{\sim}\|e_n\|^2 < \infty$$

for every orthonormal basis $\{e_n\}_{n\in\mathbb{N}}$ in $\mathcal{H}$, where $\|\cdot\|$ and ${}^{\sim}\|\cdot\|$ denote the norms corresponding to the inner products $\langle , \rangle$ and ${}^{\sim}\langle , \rangle$ respectively.

We shall need the fact that there exists an orthonormal basis in $\mathcal{H}$ which is a complete orthonormal system in $\tilde{\mathcal{H}}$, i.e. the orthogonality is preserved for this basis when considered in $\tilde{\mathcal{H}}, \tilde{\langle , \rangle}$ (we note that if $\{e_n\}_{n\in\mathbb{N}}$ is an orthonormal basis in $\mathcal{H}$, then $\{e_n\}_{n\in\mathbb{N}}$ is a total set in $\tilde{\mathcal{H}}$).

In what follows we shall keep on denoting by $\tilde{\mathcal{H}}, \tilde{\langle , \rangle}$ a Hilbert-Schmidt enlargement of $\mathcal{H}, \langle , \rangle$. We shall need the following well known results.

Lemma 1E: There exists a constant K fulfilling

$${}^{\sim}\|x\| \leq K\cdot\|x\| \quad \text{for all } x\in\mathcal{H}.$$

Lemma 2E: Let $\lambda : \mathcal{H}\times\mathcal{H} \longrightarrow \mathbb{C}$ denote a map which is conjugate linear in the first term and linear in the second. If there exists a constant K fulfilling

$$|\lambda(x,y)| \le K\cdot\|x\|\cdot\|y\| \quad \text{for all} \quad x,y\in\mathcal{H} ,$$

then there exists a linear bounded operator

$$A : \mathcal{H} \longrightarrow \mathcal{H}$$

fulfilling

$$\lambda(x,y) = \langle Ax,y\rangle \quad \text{for all} \quad x,y\in\mathcal{H} .$$

We have the following fundamental result.

Theorem 3E: There exists an orthonormal basis $\{b_n\}_{n\in\mathbb{N}}$ in $\mathcal{H}$ which is a complete orthogonal system in $\tilde{\mathcal{H}}$.

Proof: We define

$$\lambda(x,y) = \tilde{\langle}x,y\rangle : \mathcal{H}\times\mathcal{H} \longrightarrow \mathbb{C} .$$

By lemma 1E there exists a constant K fulfilling

$$|\lambda(x,y)| \le K\cdot\|x\|\cdot\|y\| \quad \text{for all} \quad x,y\in\mathcal{H}$$

by which there exists an operator $J : \mathcal{H} \longrightarrow \mathcal{H}$ with

$$\tilde{\langle}x,y\rangle = \langle Jx,y\rangle \quad \text{for all} \quad x,y\in\mathcal{H} .$$

It is easily seen that J is a strictly positive (hence self-adjoint) operator and that if $\{e_n\}_{n\in\mathbb{N}}$ denotes an orthonormal basis in $\mathcal{H}$, then

$$\begin{aligned} tr(J) &= \sum_{n=1}^{\infty} \langle e_n,Je_n\rangle = \sum_{n=1}^{\infty} \langle Je_n,e_n\rangle = \sum_{n=1}^{\infty} \tilde{\langle}e_n,e_n\rangle \\ &= \sum_{n=1}^{\infty} \tilde{\|}e_n\|^2 < \infty , \end{aligned}$$

i.e. J is a trace class operator.

Now choose an orthonormal basis $\{b_n\}_{n\in\mathbb{N}}$ in $\mathcal{H}$ consisting of eigenvectors for J with eigenvalues

$$\{\lambda_n\}_{n\in\mathbb{N}} , \quad \lambda_n > 0 .$$

We then have

1) The series $\{\lambda_n\}_{n\in\mathbb{N}}$ is convergent, i.e. $\sum_{n=1}^{\infty} \lambda_n < \infty$ since J is of trace class.

2) The orthogonality follows easily for $n,m\in\mathbb{N}$

$$\tilde{<}b_n,b_m> = <Jb_n,b_m> = \lambda_n<b_n,b_m> = \lambda_n\cdot\delta_{n,m} .$$

3) It is obvious that $\{b_n\}_{n\in\mathbb{N}}$ is total in $\tilde{\mathcal{H}},\tilde{<},>$.

■

<u>**Corollary 4E:**</u> There exists an orthonormal basis $\{b_n\}_{n\in\mathbb{N}}$ in $\mathcal{H},<,>$ and a convergent series of strictly positive numbers $\{\lambda_n\}_{n\in\mathbb{N}}$ such that

$$\tilde{\mathcal{H}} = \left\{ \sum_{n=1}^{\infty} x_n b_n \;\middle|\; \sum_{n=1}^{\infty} \lambda_n\cdot|x_n|^2 < \infty \right\} .$$

<u>**Proof:**</u> We choose $\{b_n\}_{n\in\mathbb{N}}$ and $\{\lambda_n\}_{n\in\mathbb{N}}$ as constructed in the proof of theorem 3E and the corollary follows.

■

<u>**F: The gaussian measure on Hilbert-Schmidt enlargements**</u> (cf. [23])

Using the orthonormal basis $\{b_n\}_{n\in\mathbb{N}}$ in $\mathcal{H}$ and the positive eigenvalues $\{\lambda_n\}_{n\in\mathbb{N}}$ from the above paragraph we get

$$\tilde{<}b_n,x> = <Jb_n,x> = \lambda_n<b_n,x> \quad \text{for} \quad x\in\mathcal{H} .$$

Let us define

$$\ell^2 = \left\{ \underline{x} \in \mathbb{C}^\infty \;\middle|\; \sum_{n=1}^{\infty} |x_n|^2 < \infty \right\}$$

$$\tilde{\ell}^2 = \left\{ \underline{x} \in \mathbb{C}^\infty \;\middle|\; \sum_{n=1}^{\infty} \lambda_n|x_n|^2 < \infty \right\} .$$

By identifying

$$\mathcal{H} \cong \ell^2$$
$$\tilde{\mathcal{H}} \cong \tilde{\ell}^2$$

via the orthonormal basis $\{b_n\}_{n\in\mathbb{N}}$ in $\mathcal{H}$, we can define the gaussian measure $\gamma^{\sigma}_{\mathcal{H}}$, $\sigma = \frac{1}{2}, 1$, on the Borel sets of $\tilde{\mathcal{H}}$ as the measure γ^{σ} on the corresponding Borel sets in $\tilde{\ell}^2$, and then get

$$\gamma^{\sigma}_{\mathcal{H}}(\tilde{\mathcal{H}}) = \gamma^{\sigma}(\tilde{\ell}^2) = 1 .$$

Since unitary transformations of ℓ^2 onto ℓ^2 extend to orthogonal weakly measurable linear transformations in $\tilde{\ell}^2$ and the measure γ^{σ} is invariant under these transformations, the measure $\gamma^{\sigma}_{\mathcal{H}}$ is invariant under the corresponding extensions of the unitary maps between Hilbert spaces. In particular, we have that the measure $\gamma^{\sigma}_{\mathcal{H}}$ does not depend on the chosen orthonormal basis in $\mathcal{H}$ used for identifying $\mathcal{H}$ with ℓ^2.

We would like to show that the gaussian measure does not depend on the selected Hilbert–Schmidt enlargement. This can be done by showing that if $\mathcal{H}_1$ and $\mathcal{H}_2$ denote two different Hilbert–Schmidt enlargements then there exist a Hilbert–Schmidt enlargement $\tilde{\mathcal{H}}$ fulfilling

$$\tilde{\mathcal{H}} \subset \mathcal{H}_i \qquad i=1,2 .$$

If f denotes a continuous function defined on $\mathcal{H}$ with values in $\mathbb{C}$, and f_1 and f_2 are continuous extensions of f to $\mathcal{H}_1$ and $\mathcal{H}_2$ respectively, then f_1 and f_2 are equal on $\tilde{\mathcal{H}}$, hence equal almost everywhere.

<u>**Definition 1F:**</u> Let $\mathcal{H}_1, \langle , \rangle_1$ and $\mathcal{H}_2, \langle , \rangle_2$ denote Hilbert–Schmidt enlargements of a Hilbert space $\mathcal{H}, \langle , \rangle$. We say that $\mathcal{H}_1$ is finer than $\mathcal{H}_2$ if the identity

$$I : \mathcal{H} \longrightarrow \mathcal{H}$$

extends to a continuous and one-to-one map

$$I : \mathcal{H}_1, \langle , \rangle_1 \longrightarrow \mathcal{H}_2, \langle , \rangle_2 .$$

It is obvious that if $\mathcal{H}_1$ is finer than $\mathcal{H}_2$, then there exist a constant $C > 0$ such that

$$\|x\|_2 \leq C \cdot \|x\|_1$$

for all $x \in \mathcal{H}$. The smallest C fulfilling the above is the operator norm of I in $\mathcal{B}(\mathcal{H}_1, \mathcal{H}_2)$.

Lemma 2F: Consider seminorms $\{\|\cdot\|_n\}_{n\in\mathbb{N}}$ and define $\|\cdot\|_*$ in $\mathcal{H}$ setting

$$\|x\|_*^2 = \sum_{n=1}^{\infty} \|x\|_n^2 \; .$$

If a sequence $\{x_p\}_{p\in\mathbb{N}}$ fulfills

$$\|x_p - x_q\|_* \xrightarrow[p,q]{} 0$$

and for every $n \in \mathbb{N}$

$$\|x_p\|_n \xrightarrow[p]{} 0 \; ,$$

then

$$\|x_p\|_* \xrightarrow[p]{} 0 \; .$$

Proof: The proof amounts to an adjustment of the well known method of verification that the countable direct sum of Hilbert spaces is complete.

∎

Lemma 3F: Let $\mathcal{H}_1, <,>_1$ denote a Hilbert–Schmidt enlargement of $\mathcal{H}, <,>$. If

$$\lambda = \langle a, \cdot \rangle : \mathcal{H} \longrightarrow \mathbb{C}$$

is a continuous linear functional, then there exists a Hilbert–Schmidt enlargement $\mathcal{H}_2, <,>_2$ of $\mathcal{H}, <,>$ such that

1) $\mathcal{H}_2$ is finer than $\mathcal{H}_1$

2) λ is continuous in the metric of $<,>_2$.

Proof: Use theorem 3E to choose an orthonormal basis $\{e_n\}_{n\in\mathbb{N}}$ in $\mathcal{H},<,>$ such that $\{e_n\}_{n\in\mathbb{N}}$ are pairwise orthogonal with respect to $<,>_1$. Then

$$<e_n,\cdot>_1 = \|e_n\|_1^2\cdot <e_n,\cdot> .$$

Expanding $a\in\mathcal{H}$ we get

$$a = \sum_{n=1}^{\infty} a_n\cdot e_n$$

with

$$a_n = <e_n,a>$$

and

$$\sum_{n=1}^{\infty} |a_n|^2 < \infty$$

and the functional λ can be expressed as

$$\lambda = <a,\cdot> = \sum_{n=1}^{\infty} \bar{a}_n\cdot <e_n,\cdot> .$$

Choose a strictly increasing sequence $\{m_n\}_{n\in\mathbb{N}}$ of natural numbers with $m_1 = 1$ such that

$$\sum_{n=1}^{\infty} 2^n\cdot \sum_{k=m_n}^{m_{n+1}-1} |a_k|^2 < \infty .$$

For $x,y\in\mathcal{H}$ we then define

$$<x,y>_* = \sum_{n=1}^{\infty} 2^n\cdot \overline{\left[\sum_{k=m_n}^{m_{n+1}-1} \bar{a}_k\cdot <e_k,x>\right]}\cdot \left[\sum_{k=m_n}^{m_{n+1}-1} \bar{a}_k\cdot <e_k,y>\right]$$

and

$$\|x\|_*^2 = \sum_{n=1}^{\infty} 2^n\cdot \left|\sum_{k=m_n}^{m_{n+1}-1} \bar{a}_k\cdot <e_k,x>\right|^2 .$$

Since by the Cauchy-Schwarz inequality we get

$$\|x\|_*^2 = \sum_{n=1}^{\infty} 2^n\cdot \left|\sum_{k=m_n}^{m_{n+1}-1} \bar{a}_k\cdot <e_k,x>\right|^2 = \sum_{n=1}^{\infty} 2^n\cdot \left|< \sum_{k=m_n}^{m_{n+1}-1} a_k\cdot e_k\ ,\ x >\right|^2$$
$$\leq \sum_{n=1}^{\infty} 2^n\cdot \left\| \sum_{k=m_n}^{m_{n+1}-1} a_k\cdot e_k \right\|^2\cdot \|x\|^2 = \|x\|^2\cdot \sum_{n=1}^{\infty} 2^n\cdot \sum_{k=m_n}^{m_{n+1}-1} |a_k|^2 < \infty ,$$

$\langle\cdot,\cdot\rangle_*$ is well defined on $\mathcal{H}$.

We define the inner product $\langle,\rangle_2$ in $\mathcal{H}$ setting

$$\langle x,y\rangle_2 = \langle x,y\rangle_* + \langle x,y\rangle_1 \quad \text{for } x,y\in\mathcal{H}$$

and the corresponding norm, $\|\cdot\|_2$,

$$\|x\|_2^2 = \|x\|_*^2 + \|x\|_1^2 .$$

For $x\in\mathcal{H}$ consider the expression

$$\lambda(x) = \sum_{n=1}^{\infty} \bar{a}_n\cdot\langle e_n,x\rangle$$

$$= \sum_{n=1}^{\infty} \sum_{k=m_n}^{m_{n+1}-1} \bar{a}_k\cdot\langle e_k,x\rangle = \sum_{n=1}^{\infty} (\sqrt{2})^{-n}\cdot(\sqrt{2})^n \sum_{k=m_n}^{m_{n+1}-1} \bar{a}_k\cdot\langle e_k,x\rangle .$$

Using the Cauchy–Schwarz inequality, we get

$$|\lambda(x)| \le \sum_{n=1}^{\infty} (\sqrt{2})^{-n}\cdot(\sqrt{2})^n \left| \sum_{k=m_n}^{m_{n+1}-1} \bar{a}_k\cdot\langle e_k,x\rangle \right|$$

$$\le \left[\sum_{n=1}^{\infty} 2^{-n}\right]^{\frac{1}{2}}\cdot \left[\sum_{n=1}^{\infty} 2^n \left| \sum_{k=m_n}^{m_{n+1}-1} \bar{a}_k\cdot\langle e_k,x\rangle \right|^2\right]^{\frac{1}{2}}$$

$$= \|x\|_* \le \|x\|_2 ,$$

i.e. λ is continuous in the metric of $\langle\cdot,\cdot\rangle_2$.

Since

$$\sum_{p=1}^{\infty} \|e_p\|_*^2 = \sum_{p=1}^{\infty} \sum_{n=1}^{\infty} 2^n\cdot\left| \sum_{k=m_n}^{m_{n+1}-1} \bar{a}_k\cdot\langle e_k,e_p\rangle \right|^2$$

$$= \sum_{p=1}^{\infty} \sum_{n=1}^{\infty} 2^n\cdot\left| \sum_{k=m_n}^{m_{n+1}-1} \bar{a}_k\cdot\delta_{kp} \right|^2$$

$$= \sum_{n=1}^{\infty} \sum_{p=1}^{\infty} 2^n\cdot\left| \sum_{k=m_n}^{m_{n+1}-1} \bar{a}_k\cdot\delta_{kp} \right|^2$$

$$= \sum_{n=1}^{\infty} \sum_{p=1}^{\infty} 2^n\cdot \sum_{k=m_n}^{m_{n+1}-1} |a_k|^2\cdot\delta_{kp}$$

$$\le \sum_{n=1}^{\infty} 2^n\cdot \sum_{k=m_n}^{m_{n+1}-1} |a_k|^2 < \infty ,$$

it is easy to see that

$$\sum_{p=1}^{\infty} \|e_p\|_2^2 < \infty .$$

Then the completion $\mathcal{H}_2$ of $\mathcal{H},<,>_2$ is a Hilbert–Schmidt enlargement of $\mathcal{H},<,>$.

It remains to prove that $\mathcal{H}_2$ is finer than $\mathcal{H}_1$. It is obvious that the inclusion mapping

$$I : \mathcal{H},<,>_2 \longrightarrow \mathcal{H},<,>_1$$

is continuous.

To prove that the continuous extension of I is one-to-one, it is sufficient to show that for each sequence $\{x_p\}_{p\in\mathbb{N}} \subset \mathcal{H}$ such that

1) $$\|x_p - x_q\|_2 \xrightarrow[p,q]{} 0$$

and

2) $$\|x_p\|_1 \xrightarrow[p]{} 0 ,$$

we have

$$\|x_p\|_2 \xrightarrow[p]{} 0 .$$

From 2) it follows that for every $k\in\mathbb{N}$

$$<e_k,x_p>_1 \xrightarrow[p]{} 0$$

and in particular that

$$<e_k,x_p> = \|e_k\|_1^{-2}\cdot<e_k,x_p>_1 \xrightarrow[p]{} 0$$

for every $k\in\mathbb{N}$. By linearity we get that for every $n\in\mathbb{N}$

$$2^n\cdot\left|\sum_{k=m_n}^{m_{n+1}-1} \bar{a}_k\cdot<e_k,x_p>\right|^2 \xrightarrow[p]{} 0 .$$

For $n\in\mathbb{N}$ and $x\in\mathcal{H}$ we define the seminorms $\|\cdot\|_n$,

$$\|x\|_n^2 = 2^n\cdot\left|\sum_{k=m_n}^{m_{n+1}-1} \bar{a}_k\cdot<e_k,x>\right|^2 .$$

Then by lemma 2F we get that

$$\|x_p\|_* \xrightarrow[p]{} 0$$

and thereby

$$\|x_p\|_2 \xrightarrow[p]{} 0 .$$

■

<u>**Lemma 4F:**</u> If $\{\mathcal{H}_n, <,>_n\}_{n\in\mathbb{N}}$ denotes a sequence of Hilbert–Schmidt enlargements of a Hilbert space $\mathcal{H}, <,>$ all finer than a fixed Hilbert–Schmidt enlargements $\mathcal{H}_0, <,>_0$, then there exist a Hilbert–Schmidt enlargement $\tilde{\mathcal{H}}, {}^{\sim}<,>$ finer than $\mathcal{H}_n, <,>_n$ for every $n\in\mathbb{N}$.

<u>**Proof:**</u> Take an orthonormal basis $\{e_n\}_{n\in\mathbb{N}}$ in $\mathcal{H}, <,>$. Since the inclusion mappings from $\mathcal{H}, \|\cdot\|$ to $\mathcal{H}, \|\cdot\|_n$ are Hilbert–Schmidt, we have

$$C_n = \left(\sum_{p=1}^{\infty} \|e_p\|_n^2 \right)^{\frac{1}{2}} < \infty .$$

It follows that

$$\|x\|_n \leq C_n \cdot \|x\| \quad \text{for every} \quad x\in\mathcal{H} .$$

Note that the number C_n does not depend on the choice of the basis $\{e_p\}_{p\in\mathbb{N}}$.

For $x,y\in\mathcal{H}$ we define the inner product

$${}^{\sim}<x,y> = \sum_{n=1}^{\infty} a_n \cdot <x,y>_n$$

and the norm

$${}^{\sim}\|x\|^2 = \sum_{n=1}^{\infty} a_n \cdot \|x\|_n^2 ,$$

where $\{a_n\}_{n\in\mathbb{N}}$ is a sequence of sufficiently fast decreasing numbers.

Choose

$$a_n = 2^{-n} \cdot C_n^{-2} .$$

We compute

$${}^{\sim}\|x\|^2 = \sum_{n=1}^{\infty} a_n \cdot \|x\|_n^2 \leq \sum_{n=1}^{\infty} a_n \cdot C_n^2 \cdot \|x\|^2 \leq \sum_{n=1}^{\infty} 2^{-n} \cdot \|x\|^2 \leq \|x\|^2 < \infty ,$$

and conclude that $^{\sim}\langle , \rangle$ is well defined.

Let $\tilde{\mathcal{H}}, {}^{\sim}\langle , \rangle$ denote the completion of $\mathcal{H}, {}^{\sim}\langle , \rangle$ and let $\{e_p\}_{p\in\mathbb{N}}$ be an orthonormal basis in $\mathcal{H}, \langle , \rangle$. Then we have

$$\sum_{p=1}^{\infty} {}^{\sim}\|e_p\|^2 = \sum_{p=1}^{\infty} \sum_{n=1}^{\infty} a_n \cdot \|e_p\|_n^2$$

$$= \sum_{n=1}^{\infty} a_n \sum_{p=1}^{\infty} \|e_p\|_n^2 = \sum_{n=1}^{\infty} a_n \cdot C_n^2 = \sum_{n=1}^{\infty} 2^{-n} < \infty .$$

We have to prove that $\tilde{\mathcal{H}}$ is a Hilbert–Schmidt enlargement of $\mathcal{H}$. Take $n\in\mathbb{N}$. We must show that $\tilde{\mathcal{H}}$ is finer than $\mathcal{H}_n$. It is obvious that the inclusion mapping

$$I : \mathcal{H}, {}^{\sim}\langle , \rangle \longrightarrow \mathcal{H}_n, \langle , \rangle_n$$

is continuous.

It remains to prove that the extension of I over $\tilde{\mathcal{H}}$ is one–to–one. Assume that $\{x_p\}_{p\in\mathbb{N}} \subset \mathcal{H}$ is a Cauchy–sequence in the $^{\sim}\|\cdot\|$ – norm and that $\|x_p\|_0 \xrightarrow[p]{} 0$. Then for every $n\in\mathbb{N}$

$$\|x_p - x_q\|_n \xrightarrow[p,q]{} 0 .$$

We know that for every n the enlargement $\mathcal{H}_n$ is finer than $\mathcal{H}_0$ and hence $\|x_p\|_0 \xrightarrow[p]{} 0$ implies

$$\|x_p\|_n \xrightarrow[p]{} 0 .$$

Since

$$^{\sim}\|x_p\| = \left(\sum_{n=1}^{\infty} a_n \cdot \|x_p\|_n^2 \right)^{\frac{1}{2}}$$

we get by lemma 2F that

$$^{\sim}\|x_p\| \xrightarrow[p]{} 0$$

which proves that the inclusion mapping of $\tilde{\mathcal{H}}$ into $\mathcal{H}_0$ is one-to-one. Since the inclusion mapping of $\mathcal{H}_n$ into $\mathcal{H}_0$ is one-to-one, the injection of $\tilde{\mathcal{H}}$ into $\mathcal{H}_n$ must be one-to-one as well.

■

We are ready to prove the announced principal result.

Theorem 5F: To every pair of Hilbert–Schmidt enlargements of a Hilbert space there always exists a Hilbert–Schmidt enlargement that is finer than any enlargements from the pair.

Proof: Let $\mathcal{H}_1, \langle,\rangle_1$ and $\mathcal{H}_2, \langle,\rangle_2$ be a pair of Hilbert–Schmidt enlargements of a Hilbert space $\mathcal{H}, \langle,\rangle$. Choose an orthonormal basis $\{e_n\}_{n\in\mathbb{N}}$ in $\mathcal{H}, \langle,\rangle$ that is a complete orthogonal system in $\mathcal{H}_1, \langle,\rangle_1$. Normalize $\{e_n\}_{n\in\mathbb{N}}$ in $\mathcal{H}_1, \langle,\rangle_1$ by setting

$$b_n = \|e_n\|_1^{-1} \cdot e_n$$

and for every n define the continuous linear functional λ_n in $\mathcal{H}_1, \langle,\rangle_1$ by

$$\lambda_n = \langle b_n, \cdot\rangle_1 \ .$$

Note that λ_n is continuous in $\mathcal{H}, \langle,\rangle$ and hence

$$\lambda_n = \|e_n\|_1 \cdot \langle e_n, \cdot\rangle \quad \text{in } \mathcal{H} \ .$$

By lemma 3F and lemma 4F we find a Hilbert–Schmidt enlargement $\mathcal{H}_3, \langle,\rangle_3$ of $\mathcal{H}, \langle,\rangle$ that is finer than $\mathcal{H}_2, \langle,\rangle_2$ and such that the functionals $\{\lambda_n\}_{n\in\mathbb{N}}$ are continuous in $\mathcal{H}_3, \langle,\rangle_3$.

For $x, y \in \mathcal{H}$ we define the inner product

$$\sim\langle x,y\rangle = \langle x,y\rangle_1 + \langle x,y\rangle_3$$

with the corresponding norm

$$\sim\|x\| = (\|x\|_1^2 + \|x\|_3^2)^{\frac{1}{2}} \ .$$

Then we define $\tilde{\mathcal{H}}$ to be the completion of $\mathcal{H}, \sim\langle,\rangle$. We have

$$\sum_{n=1}^{\infty} \sim\|e_n\|^2 = \sum_{n=1}^{\infty} \|e_n\|_1^2 + \sum_{n=1}^{\infty} \|e_n\|_3^2 < \infty \ ,$$

which makes the inner product well defined.

The inclusion mappings

$$I : \mathcal{H}, \sim\langle,\rangle \longrightarrow \mathcal{H}_1, \langle,\rangle_1$$

and

$$I : \mathcal{H}, \sim\langle,\rangle \longrightarrow \mathcal{H}_3, \langle,\rangle_3$$

are obviously continuous.

That their extensions are one-to-one follows from the fact that the functionals $\{\lambda_n\}_{n\in\mathbb{N}}$ are continuous in both the $\|\cdot\|_1$ – norm and the $\|\cdot\|_3$ – norm and that these functionals are dense in $(\mathcal{H},<,>)'$ and hence dense in both $(\mathcal{H}_1,<,>_1)'$ and $(\mathcal{H}_3,<,>_3)'$.

■

When considering gaussian measures we always let $\mathcal{H},<,>$ denote a complex Hilbert space and $H,<,>$ a real Hilbert space.

The measure $\gamma_{\mathcal{H}}^{\frac{1}{2}}$ will then denote the measure in $\tilde{l}^2 \subset \mathbb{C}^\infty$ given on $\mathbb{C}^n$ by

$$\pi^{-n}\cdot\exp(-(|z_1|^2 + \dots + |z_n|^2))d\underline{z} ,$$

where $\underline{z} = (z_1,z_2,\dots,z_n) \in \mathbb{C}^n$ with $z_k = x_k + i\cdot y_k \in \mathbb{C}$

and $d\underline{z} = dx_1dy_1dx_2dy_2\dots dx_ndy_n$ indicates the Lebesgue integration in $\mathbb{R}^{2n}$.

In a similar way γ_H denotes the measure in $\tilde{l}^2 \subset \mathbb{R}^\infty$ given on $\mathbb{R}^n$ by

$$(2\pi)^{-\frac{1}{2}n}\cdot\exp(-\tfrac{1}{2}(x_1^2 + \dots + x_n^2))d\underline{x} ,$$

where $\underline{x} = (x_1,x_2,\dots,x_n) \in \mathbb{R}^n$

and $d\underline{x} = dx_1dx_2\dots dx_n$ indicates the Lebesgue integration in $\mathbb{R}^n$.

Observe that the measure $\gamma_{\mathcal{H}}^{\frac{1}{2}}$ on a complex Hilbert space $\mathcal{H}$ is produced by considering $\mathcal{H}$ as a real Hilbert space with inner product

$$<\cdot,\cdot>_{\mathbb{R}} = \tfrac{1}{2}(<\cdot,\cdot> + \overline{<\cdot,\cdot>}) = \mathrm{Re}\ <\cdot,\cdot> .$$

References

[1] V. Bargmann, On a Hilbert Space of Analytic Functions and an Associated Integral Transform Part I, Pure Appl. Math. Sci. 14 (1961), page 187–214.

[2] O. Bratteli and D. W. Robinson, Operator Algebras and Quantum Statistical Mechanics II, Springer Verlag, (1981).

[3] J. Cook, The mathematics of second quantisation, Trans. Amer. Math. Soc. 74 (1953), page 222–245.

[4] W. Feller, Acknowledgment, Proc. Nat. Acad. Sci. USA, Vol 48 (1962) page 2204.

[5] R.J. Glauber, Coherent and Incoherent States of the Radiation Field, Physical Review 131(1963), page 2766–2788.

[6] Alain Guichardet, Symmetric Hilbert Spaces and Related Topics, Lecture Notes in Mathematics 261, Springer Verlag

[7] R.L. Hudson and J.M. Lindsay, A non–commutative martingale representation theorem for non–Fock Quantum Brownian Motion, Journal of Functional Analysis 61 (1985), page 202–221.

[8] R.L. Hudson and K.R. Parthasarathy, Quantum Ito's Formula and Stochastic Evolutions, Commun. Math. Phys. 93 (1984), page 301–323.

[9] R.L. Hudson and R.F. Streater, Ito's formula is the chain rule with Wick ordering, Physics Letters, vol 86A, number 5, page 277–279.

[10] P. Kristensen, L. Mejlbo, E.T. Poulsen, Tempered distributions in infinitely many dimensions I,II,III

I : Commun. Math .Phys. 1(1965) page 175–214.
II : Math. Scand. 14(1964) page 129–150.
III : Commun. Math. Phys. 6(1967) page 29– 48.

[11] W.H. Louisell, Quantum Statistical Properties of Radiation, Wiley 1973.

[12] J.E. Moyal, Quantum Mechanics as a Statistical Theory, Proceedings of the Cambridge Philosophical Society 45 (1949) page 99–124.

[13] Klaus Mølmer and Wojtek Słowikowski, A new Hilbert space approach to the multimode squeezing of light, J.Phys. A:Math.Gen 21 (1988) page 2565–2571.

[14] E. Nelson, The free Markoff field, J. Funct. Anal. 12(1973) page 211–227.

[15] I.E. Segal, Tensor algebras over Hilbert space I, Trans. Amer. Math. Soc. 81 (1956), page 106–134.

[16] I.E. Segal, The complex-wave representation of the free Boson field, Topics in Functional Analysis, Adv. in Math., Supplementary Studies Vol 3 (1978), page 321–345.

[17] I.E. Segal, J.C Baez and Zhengfang Zhou, Introduction to algebraic and constructive quantum field theory, Princeton Univ. Press, to appear.

[18] G.E. Silov and Fan Dyk Tin, Integral, Measure and Derivative, Nauka, Moscow (1967), (Russian).

[19] B. Simon, The $P(\Phi)_2$ Euclidean (Quantum) Field Theory, Princeton University Press (1974).

[20] W. Słowikowski, Ultracoherence in Bose algebras, Advances in Applied Mathematics 9 (1988) 377-427.

[21] W. Słowikowski, Bose algebras of operators and the Wigner-Weyl theory, to appear.

[22] W. Słowikowski, Quantization of Questionnaires, Math. Comput. Modelling. Vol 10 No. 9 (1988), page 697–709.

[23] W. Słowikowski, Concerning pre-supports of linear probability measures, Lecture Notes in Mathematics 794, Proceedings Measure Theory Oberwolfach 1979, Springer.

[24] W. Słowikowski, Measure Theory Applications to Stochastic Analysis, Proceedings, Oberwolfach, Germany 1977, Lecture Notes in Math. vol 695, page 179–191, Springer 1978.

[25] A. Unterberger, Les operatéurs metadifferétiels, Lecture Notes in Physics 126 (1980), page 205–241.

[26] L.V. Zyla and Rui J. P. deFigueiredo, Nonlinear System Identification based on a Fock space framework, SIAM J. Control and Optimization, Vol 21 NO. 6 (November 1983), page 931–939.

Subject index

annihilation operator 1,3,6,8,10, 21,25,61,64,79,81,83,85,92
anti-commutator 1
anti-normal 92
base space 1,2,4,10,23,65,83,85, 87
Bose albebra 1,2,3,4,6,10,11,20, 22,23,65,79,83,84,85,88
- extende 36,65,83
- Fock space 1,2,3
Campbell-Baker-Hausdorff 50,51, 59,91,92
coherent 2,33,34,43,44,63,64,66, 83,89,92
commutation 9,10,11,12,45,55,59, 60,65,82,85,93
complex wave representation 2,53, 66,69,70,71,78,80,87,90
conjugation 1,2,3,53,54,55,57,69, 73,74,76,77,80,84,88,89,90
creation operator 1,3,6,8,10,21, 25,64,79,81,83,85,92
cylinder set 96,97,101,107
derivation 1,6,10,11,30
Fourier transformation 25,31,32,71
free commutative algebra 1,4
- product 1,4,18
gaussian content 96
- measure 3,68,76,78,80,86,96,99, 107,115,116,117,118,120,121,122
Halmos 27,94
Heisenberg 33,43
Hermite 2,31,32
Hilbert-Schmidt enlargement 68,71,76, 80,86,87,90,91,118,120,121,122, 125,126,127,128
Kolmogorov extension theorem 97
- inequality 103,105
- large number theorem 104
- zero one law 102,111
Leibniz rule 6,30,31,36,37,39,61,82
linear measurable functional 102,103, 106,108,109,110,111,113,114,115, 117,122,123,128,129
Nelson 27
one-parameter group 29
ψ- picture 65,84,88
- product 59,61,62,63,64,65,78,88
- value 72,90
real wave representation 2,3,53, 72,77,90,91
second quantization 30
Stone 29,30
tame 100,101,102
total 16,53,66,67,92,118,120
ultracoherent 57
vacuum 1,4,10,20,23,65,83,85,92
value of 66,90
weak measurable linear trans- formation 115,116,121
Weyl 1,45,52,91
Wick 22,45,59,79,81,83,88
Wiener 16,20
Wigner 3,86

Lecture Notes in Mathematics

For information about Vols. 1–1272
please contact your bookseller or Springer-Verlag

Vol. 1273: G.-M. Greuel, G. Trautmann (Eds.), Singularities, Representation of Algebras, and Vector Bundles. Proceedings, 1985. XIV, 383 pages. 1987.

Vol. 1274: N.C. Phillips, Equivariant K-Theory and Freeness of Group Actions on C*-Algebras. VIII, 371 pages. 1987.

Vol. 1275: C.A. Berenstein (Ed.), Complex Analysis I. Proceedings, 1985–86. XV, 331 pages. 1987.

Vol. 1276: C.A. Berenstein (Ed.), Complex Analysis II. Proceedings, 1985–86. IX, 320 pages. 1987.

Vol. 1277: C.A. Berenstein (Ed.), Complex Analysis III. Proceedings, 1985–86. X, 350 pages 1987.

Vol. 1278: S.S. Koh (Ed.), Invariant Theory. Proceedings, 1985. V, 102 pages. 1987.

Vol. 1279: D. Iesan, Saint Venant's Problem. VIII, 162 Seiten. 1987.

Vol. 1280: E. Neher, Jordan Triple Systems by the Grid Approach. XII, 193 pages. 1987.

Vol. 1281: O.H. Kegel, F. Menegazzo, G. Zacher (Eds.), Group Theory. Proceedings, 1986. VII, 179 pages. 1987.

Vol. 1282: D.E. Handelman, Positive Polynomials, Convex Integral Polytopes, and a Random Walk Problem. XI, 136 pages. 1987.

Vol. 1283: S. Mardesic´, J. Segal (Eds.), Geometric Topology and Shape Theory. Proceedings, 1986. V, 261 pages. 1987.

Vol. 1284: B.H. Matzat, Konstruktive Galoistheorie. X, 286 pages. 1987.

Vol. 1285: I.W. Knowles, Y. Saito (Eds.), Differential Equations and Mathematical Physics. Proceedings, 1986. XVI, 499 pages. 1987.

Vol. 1286: H.R. Miller, D.C. Ravenel (Eds.), Algebraic Topology. Proceedings, 1986. VII, 341 pages. 1987.

Vol. 1287: E.B. Saff (Ed.), Approximation Theory, Tampa, Proceedings, 1985–1986. V, 228 pages. 1987.

Vol. 1288: Yu.L. Rodin, Generalized Analytic Functions on Riemann Surfaces. V, 128 pages, 1987.

Vol. 1289: Yu.I. Manin (Ed.), K-Theory, Arithmetic and Geometry. Seminar, 1984–1986. V, 399 pages. 1987.

Vol. 1290: G. Wüstholz (Ed.), Diophantine Approximation and Transcendence Theory. Seminar, 1985. V, 243 pages. 1987.

Vol. 1291: C. Mœglin, M.-F. Vignéras, J.-L. Waldspurger, Correspondances de Howe sur un Corps p-adique. VII, 163 pages. 1987.

Vol. 1292: J.T. Baldwin (Ed.), Classification Theory. Proceedings, 1985. VI, 500 pages. 1987.

Vol. 1293: W. Ebeling, The Monodromy Groups of Isolated Singularities of Complete Intersections. XIV, 153 pages. 1987.

Vol. 1294: M. Queffélec, Substitution Dynamical Systems – Spectral Analysis. XIII, 240 pages. 1987.

Vol. 1295: P. Lelong, P. Dolbeault, H. Skoda (Réd.), Séminaire d'Analyse P. Lelong – P. Dolbeault – H. Skoda. Seminar. 1985/1986. VII, 283 pages. 1987.

Vol. 1296: M.-P. Malliavin (Ed.), Séminaire d'Algèbre Paul Dubreil et Marie-Paule Malliavin. Proceedings, 1986. IV, 324 pages. 1987.

Vol. 1297: Zhu Y.-l., Guo B.-y. (Eds.), Numerical Methods for Partial Differential Equations. Proceedings. XI, 244 pages. 1987.

Vol. 1298: J. Aguadé, R. Kane (Eds.), Algebraic Topology, Barcelona 1986. Proceedings. X, 255 pages. 1987.

Vol. 1299: S. Watanabe, Yu.V. Prokhorov (Eds.), Probability Theory and Mathematical Statistics. Proceedings, 1986. VIII, 589 pages. 1988.

Vol. 1300: G.B. Seligman, Constructions of Lie Algebras and their Modules. VI, 190 pages. 1988.

Vol. 1301: N. Schappacher, Periods of Hecke Characters. XV, 160 pages. 1988.

Vol. 1302: M. Cwikel, J. Peetre, Y. Sagher, H. Wallin (Eds.), Function Spaces and Applications. Proceedings, 1986. VI, 445 pages. 1988.

Vol. 1303: L. Accardi, W. von Waldenfels (Eds.), Quantum Probability and Applications III. Proceedings, 1987. VI, 373 pages. 1988.

Vol. 1304: F.Q. Gouvêa, Arithmetic of p-adic Modular Forms. VIII, 121 pages. 1988.

Vol. 1305: D.S. Lubinsky, E.B. Saff, Strong Asymptotics for Extremal Polynomials Associated with Weights on R. VII, 153 pages. 1988.

Vol. 1306: S.S. Chern (Ed.), Partial Differential Equations. Proceedings, 1986. VI, 294 pages. 1988.

Vol. 1307: T. Murai, A Real Variable Method for the Cauchy Transform, and Analytic Capacity. VIII, 133 pages. 1988.

Vol. 1308: P. Imkeller, Two-Parameter Martingales and Their Quadratic Variation. IV, 177 pages. 1988.

Vol. 1309: B. Fiedler, Global Bifurcation of Periodic Solutions with Symmetry. VIII, 144 pages. 1988.

Vol. 1310: O.A. Laudal, G. Pfister: Local Moduli and Singularities. V, 117 pages. 1988.

Vol. 1311: A. Holme, R. Speiser (Eds.), Algebraic Geometry, Sundance 1986. Proceedings. VI, 320 pages. 1988.

Vol. 1312: N.A. Shirokov, Analytic Functions Smooth up to the Boundary. III, 213 pages. 1988.

Vol. 1313: F. Colonius. Optimal Periodic Control. VI, 177 pages. 1988.

Vol. 1314: A. Futaki. Kähler-Einstein Metrics and Integral Invariants. IV, 140 pages. 1988.

Vol. 1315: R.A. McCoy, I. Ntantu, Topological Properties of Spaces of Continuous, Functions. IV, 124 pages. 1988.

Vol. 1316: H. Korezlioglu, A.S. Ustunel (Eds.), Stochastic Analysis and Related Topics. Proceedings, 1986. V, 371 pages. 1988.

Vol. 1317: J. Lindenstrauss, V.D. Milman (Eds.), Geometric Aspects of Functional Analysis. Seminar, 1986–87. VII, 289 pages. 1988.

Vol. 1318: Y. Felix (Ed.), Algebraic Topology – Rational Homotopy. Proceedings, 1986. VIII, 245 pages. 1988.

Vol. 1319: M. Vuorinen, Conformal Geometry and Quasiregular Mappings. XIX, 209 pages. 1988.

Vol. 1320: H. Jürgensen, G. Lallement, H.J. Weinert (Eds.), Semigroups, Theory and Applications. Proceedings, 1986. X, 416 pages. 1988.

Vol. 1321: J. Azéma, P.A. Meyer, M. Yor (Eds.), Séminaire de Probabilités XXII. Proceedings. IV, 600 pages. 1988.

Vol. 1322: M. Métivier, S. Watanabe (Eds.), Stochastic Analysis. Proceedings, 1987. VII, 197 pages. 1988.

Vol. 1323: D.R. Anderson, H.J. Munkholm, Boundedly Controlled Topology. XII, 309 pages. 1988.

Vol. 1324: F. Cardoso, D.G. de Figueiredo, R. Iório, O. Lopes (Eds.), Partial Differential Equations. Proceedings, 1986. VIII, 433 pages. 1988.

Vol. 1325: A. Truman, I.M. Davies (Eds.), Stochastic Mechanics and Stochastic Processes. Proceedings, 1986. V, 220 pages. 1988.

Vol. 1326: P.S. Landweber (Ed.), Elliptic Curves and Modular Forms in Algebraic Topology. Proceedings, 1986. V, 224 pages. 1988.

Vol. 1327: W. Bruns, U. Vetter, Determinantal Rings. VII, 236 pages. 1988.

Vol. 1328: J.L. Bueso, P. Jara, B. Torrecillas (Eds.), Ring Theory. Proceedings, 1986. IX, 331 pages. 1988.

Vol. 1329: M. Alfaro, J.S. Dehesa, F.J. Marcellan, J.L. Rubio de Francia, J. Vinuesa (Eds.): Orthogonal Polynomials and their Applications. Proceedings, 1986. XV, 334 pages. 1988.

Vol. 1330: A. Ambrosetti, F. Gori, R. Lucchetti (Eds.), Mathematical Economics. Montecatini Terme 1986. Seminar. VII, 137 pages. 1988.

Vol. 1331: R. Bamón, R. Labarca, J. Palis Jr. (Eds.), Dynamical Systems, Valparaiso 1986. Proceedings. VI, 250 pages. 1988.

Vol. 1332: E. Odell, H. Rosenthal (Eds.), Functional Analysis. Proceedings. 1986–87. V, 202 pages. 1988.

Vol. 1333: A.S. Kechris, D.A. Martin, J.R. Steel (Eds.), Cabal Seminar 81–85. Proceedings, 1981–85. V, 224 pages. 1988.

Vol. 1334: Yu.G. Borisovich, Yu.E. Gliklikh (Eds.), Global Analysis – Studies and Applications III. V, 331 pages. 1988.

Vol. 1335: F. Guillén, V. Navarro Aznar, P. Pascual-Gainza, F. Puerta, Hyperrésolutions cubiques et descente cohomologique. XII, 192 pages. 1988.

Vol. 1336: B. Helffer, Semi-Classical Analysis for the Schrödinger Operator and Applications. V, 107 pages. 1988.

Vol. 1337: E. Sernesi (Ed.), Theory of Moduli. Seminar, 1985. VIII, 232 pages. 1988.

Vol. 1338: A.B. Mingarelli, S.G. Halvorsen. Non-Oscillation Domains of Differential Equations with Two Parameters. XI, 109 pages. 1988.

Vol. 1339: T. Sunada (Ed.), Geometry and Analysis of Manifolds. Proceedings, 1987. IX, 277 pages. 1988.

Vol. 1340: S. Hildebrandt, D.S. Kinderlehrer, M. Miranda (Eds.), Calculus of Variations and Partial Differential Equations. Proceedings, 1986. IX, 301 pages. 1988.

Vol. 1341: M. Dauge, Elliptic Boundary Value Problems on Corner Domains. VIII. 259 pages. 1988.

Vol. 1342: J.C. Alexander (Ed.), Dynamical Systems. Proceedings, 1986–87. VIII, 726 pages. 1988.

Vol. 1343: H. Ulrich, Fixed Point Theory of Parametrized Equivariant Maps. VII, 147 pages. 1988.

Vol. 1344: J. Král, J. Lukes, J. Netuka, J. Vesely´ (Eds.), Potential Theory – Surveys and Problems. Proceedings, 1987. VIII, 271 pages. 1988.

Vol. 1345: X. Gomez-Mont, J. Seade, A. Verjovski (Eds.), Holomorphic Dynamics. Proceedings, 1986. VII. 321 pages. 1988.

Vol. 1346: O.Ya. Viro (Ed.), Topology and Geometry – Rohlin Seminar. XI, 581 pages. 1988.

Vol. 1347: C. Preston, Iterates of Piecewise Monotone Mappings on an Interval. V, 166 pages. 1988.

Vol. 1348: F. Borceux (Ed.), Categorical Algebra and its Applications. Proceedings, 1987. VIII, 375 pages. 1988.

Vol. 1349: E. Novak, Deterministic and Stochastic Error Bounds in Numerical Analysis. V, 113 pages. 1988.

Vol. 1350: U. Koschorke (Ed.), Differential Topology Proceedings, 1987, VI, 269 pages. 1988.

Vol. 1351: I. Laine, S. Rickman, T. Sorvali (Eds.), Complex Analysis, Joensuu 1987. Proceedings. XV, 378 pages. 1988.

Vol. 1352: L.L. Avramov, K.B. Tchakerian (Eds.), Algebra – Some Current Trends. Proceedings. 1986. IX, 240 Seiten. 1988.

Vol. 1353: R.S. Palais, Ch.-l. Teng, Critical Point Theory and Submanifold Geometry. X, 272 pages. 1988.

Vol. 1354: A. Gómez, F. Guerra, M.A. Jiménez, G. López (Eds.), Approximation and Optimization. Proceedings, 1987. VI, 280 pages. 1988.

Vol. 1355: J. Bokowski, B. Sturmfels, Computational Synthetic Geometry. V, 168 pages. 1989.

Vol. 1356: H. Volkmer, Multiparameter Eigenvalue Problems and Expansion Theorems. VI, 157 pages. 1988.

Vol. 1357: S. Hildebrandt, R. Leis (Eds.), Partial Differential Equations and Calculus of Variations. VI, 423 pages. 1988.

Vol. 1358: D. Mumford, The Red Book of Varieties and Schemes. V, 309 pages. 1988.

Vol. 1359: P. Eymard, J.-P. Pier (Eds.) Harmonic Analysis. Proceedings, 1987. VIII, 287 pages. 1988.

Vol. 1360: G. Anderson, C. Greengard (Eds.), Vortex Methods. Proceedings, 1987. V, 141 pages. 1988.

Vol. 1361: T. tom Dieck (Ed.), Algebraic Topology and Transformation Groups. Proceedings. 1987. VI, 298 pages. 1988.

Vol. 1362: P. Diaconis, D. Elworthy, H. Föllmer, E. Nelson, G.C. Papanicolaou, S.R.S. Varadhan. École d´ Été de Probabilités de Saint-Flour XV–XVII. 1985–87 Editor: P.L. Hennequin. V, 459 pages. 1988.

Vol. 1363: P.G. Casazza, T.J. Shura, Tsirelson´s Space. VIII, 204 pages. 1988.

Vol. 1364: R.R. Phelps, Convex Functions, Monotone Operators and Differentiability. IX, 115 pages. 1989.

Vol. 1365: M. Giaquinta (Ed.), Topics in Calculus of Variations. Seminar, 1987. X, 196 pages. 1989.

Vol. 1366: N. Levitt, Grassmannians and Gauss Maps in PL-Topology. V, 203 pages. 1989.

Vol. 1367: M. Knebusch, Weakly Semialgebraic Spaces. XX, 376 pages. 1989.

Vol. 1368: R. Hübl, Traces of Differential Forms and Hochschild Homology. III, 111 pages. 1989.

Vol. 1369: B. Jiang, Ch.-K. Peng, Z. Hou (Eds.), Differential Geometry and Topology. Proceedings, 1986–87. VI, 366 pages. 1989.

Vol. 1370: G. Carlsson, R.L. Cohen, H.R. Miller, D.C. Ravenel (Eds.), Algebraic Topology. Proceedings, 1986. IX, 456 pages. 1989.

Vol. 1371: S. Glaz, Commutative Coherent Rings. XI, 347 pages. 1989.

Vol. 1372: J. Azéma, P.A. Meyer, M. Yor (Eds.), Séminaire de Probabilités XXIII. Proceedings. IV, 583 pages. 1989.

Vol. 1373: G. Benkart, J.M. Osborn (Eds.), Lie Algebras. Madison 1987. Proceedings. V, 145 pages. 1989.

Vol. 1374: R.C. Kirby, The Topology of 4-Manifolds. VI, 108 pages. 1989.

Vol. 1375: K. Kawakubo (Ed.), Transformation Groups. Proceedings, 1987. VIII, 394 pages, 1989.

Vol. 1376: J. Lindenstrauss, V.D. Milman (Eds.), Geometric Aspects of Functional Analysis. Seminar (GAFA) 1987–88. VII, 288 pages. 1989.

Vol. 1377: J.F. Pierce, Singularity Theory, Rod Theory, and Symmetry-Breaking Loads. IV, 177 pages. 1989.

Vol. 1378: R.S. Rumely, Capacity Theory on Algebraic Curves. III, 437 pages. 1989.

Vol. 1379: H. Heyer (Ed.), Probability Measures on Groups IX. Proceedings, 1988. VIII, 437 pages. 1989.

Vol. 1380: H.P. Schlickewei, E. Wirsing (Eds.), Number Theory, Ulm 1987. Proceedings. V, 266 pages. 1989.

Vol. 1381: J.-O. Strömberg, A. Torchinsky, Weighted Hardy Spaces. V, 193 pages. 1989.

Vol. 1382: H. Reiter, Metaplectic Groups and Segal Algebras. XI, 128 pages. 1989.

Vol. 1383: D.V. Chudnovsky, G.V. Chudnovsky, H. Cohn, M.B. Nathanson (Eds.), Number Theory, New York 1985–88. Seminar. V, 256 pages. 1989.

Vol. 1384: J. Garcia-Cuerva (Ed.), Harmonic Analysis and Partial Differential Equations. Proceedings, 1987. VII, 213 pages. 1989.

Vol. 1385: A.M. Anile, Y. Choquet-Bruhat (Eds.), Relativistic Fluid Dynamics. Seminar, 1987. V, 308 pages. 1989.

Vol. 1386: A. Bellen, C.W. Gear, E. Russo (Eds.), Numerical Methods for Ordinary Differential Equations. Proceedings, 1987. VII, 136 pages. 1989.

Vol. 1387: M. Petkovi´c, Iterative Methods for Simultaneous Inclusion of Polynomial Zeros. X, 263 pages. 1989.

Vol. 1388: J. Shinoda, T.A. Slaman, T. Tugué (Eds.), Mathematical Logic and Applications. Proceedings, 1987. V, 223 pages. 1989.

Vol. 1000: Second Edition. H. Hopf, Differential Geometry in the Large. VII, 184 pages. 1989.

Vol. 1389: E. Ballico, C. Ciliberto (Eds.), Algebraic Curves and Projective Geometry. Proceedings, 1988. V, 288 pages. 1989.

Vol. 1390: G. Da Prato, L. Tubaro (Eds.), Stochastic Partial Differential Equations and Applications II. Proceedings, 1988. VI, 258 pages. 1989.

Vol. 1391: S. Cambanis, A. Weron (Eds.), Probability Theory on Vector Spaces IV. Proceedings, 1987. VIII, 424 pages. 1989.

Vol. 1392: R. Silhol, Real Algebraic Surfaces. X, 215 pages. 1989.

Vol. 1393: N. Bouleau, D. Feyel, F. Hirsch, G. Mokobodzki (Eds.), Séminaire de Théorie du Potentiel Paris, No. 9. Proceedings. VI, 265 pages. 1989.

Vol. 1394: T.L. Gill, W.W. Zachary (Eds.), Nonlinear Semigroups, Partial Differential Equations and Attractors. Proceedings, 1987. IX, 233 pages. 1989.

Vol. 1395: K. Alladi (Ed.), Number Theory, Madras 1987. Proceedings. VII, 234 pages. 1989.

Vol. 1396: L. Accardi, W. von Waldenfels (Eds.), Quantum Probability and Applications IV. Proceedings, 1987. VI, 355 pages. 1989.

Vol. 1397: P.R. Turner (Ed.), Numerical Analysis and Parallel Processing. Seminar, 1987. VI, 264 pages. 1989.

Vol. 1398: A.C. Kim, B.H. Neumann (Eds.), Groups – Korea 1988. Proceedings. V, 189 pages. 1989.

Vol. 1399: W.-P. Barth, H. Lange (Eds.), Arithmetic of Complex Manifolds. Proceedings, 1988. V, 171 pages. 1989.

Vol. 1400: U. Jannsen. Mixed Motives and Algebraic K-Theory. XIII, 246 pages. 1990.

Vol. 1401: J. Steprans, S. Watson (Eds.), Set Theory and its Applications. Proceedings, 1987. V, 227 pages. 1989.

Vol. 1402: C. Carasso, P. Charrier, B. Hanouzet, J.-L. Joly (Eds.), Nonlinear Hyperbolic Problems. Proceedings, 1988. V, 249 pages. 1989.

Vol. 1403: B. Simeone (Ed.), Combinatorial Optimization. Seminar, 1986. V, 314 pages. 1989.

Vol. 1404: M.-P. Malliavin (Ed.), Séminaire d´Algèbre Paul Dubreil et Marie-Paul Malliavin. Proceedings, 1987–1988. IV, 410 pages. 1989.

Vol. 1405: S. Dolecki (Ed.), Optimization. Proceedings, 1988. V, 223 pages. 1989. Vol. 1406: L. Jacobsen (Ed.), Analytic Theory of Continued Fractions III. Proceedings, 1988. VI, 142 pages. 1989.

Vol. 1407: W. Pohlers, Proof Theory. VI, 213 pages. 1989.

Vol. 1408: W. Lück, Transformation Groups and Algebraic K-Theory. XII, 443 pages. 1989.

Vol. 1409: E. Hairer, Ch. Lubich, M. Roche. The Numerical Solution of Differential-Algebraic Systems by Runge-Kutta Methods. VII, 139 pages. 1989.

Vol. 1410: F.J. Carreras, O. Gil-Medrano, A.M. Naveira (Eds.), Differential Geometry. Proceedings, 1988. V, 308 pages. 1989.

Vol. 1411: B. Jiang (Ed.), Topological Fixed Point Theory and Applications. Proceedings. 1988. VI, 203 pages. 1989.

Vol. 1412: V.V. Kalashnikov, V.M. Zolotarev (Eds.), Stability Problems for Stochastic Models. Proceedings, 1987. X, 380 pages. 1989.

Vol. 1413: S. Wright, Uniqueness of the Injective III_1 Factor. III, 108 pages. 1989.

Vol. 1414: E. Ramirez de Arellano (Ed.), Algebraic Geometry and Complex Analysis. Proceedings, 1987. VI, 180 pages. 1989.

Vol. 1415: M. Langevin, M. Waldschmidt (Eds.), Cinquante Ans de Polynômes. Fifty Years of Polynomials. Proceedings, 1988. IX, 235 pages.1990.

Vol. 1416: C. Albert (Ed.), Géométrie Symplectique et Mécanique. Proceedings, 1988. V, 289 pages. 1990.

Vol. 1417: A.J. Sommese, A. Biancofiore, E.L. Livorni (Eds.), Algebraic Geometry. Proceedings, 1988. V, 320 pages. 1990.

Vol. 1418: M. Mimura (Ed.), Homotopy Theory and Related Topics. Proceedings, 1988. V, 241 pages. 1990.

Vol. 1419: P.S. Bullen, P.Y. Lee, J.L. Mawhin, P. Muldowney, W.F. Pfeffer (Eds.), New Integrals. Proceedings, 1988. V, 202 pages. 1990.

Vol. 1420: M. Galbiati, A. Tognoli (Eds.), Real Analytic Geometry. Proceedings, 1988. IV, 366 pages. 1990.

Vol. 1421: H.A. Biagioni, A Nonlinear Theory of Generalized Functions, XII, 214 pages. 1990.

Vol. 1422: V. Villani (Ed.), Complex Geometry and Analysis. Proceedings, 1988. V, 109 pages. 1990.

Vol. 1423: S.O. Kochman, Stable Homotopy Groups of Spheres: A Computer-Assisted Approach. VIII, 330 pages. 1990.

Vol. 1424: F.E. Burstall, J.H. Rawnsley, Twistor Theory for Riemannian Symmetric Spaces. III, 112 pages. 1990.

Vol. 1425: R.A. Piccinini (Ed.), Groups of Self-Equivalences and Related Topics. Proceedings, 1988. V, 214 pages. 1990.

Vol. 1426: J. Azéma, P.A. Meyer, M. Yor (Eds.), Séminaire de Probabilités XXIV, 1988/89. V, 490 pages. 1990.

Vol. 1427: A. Ancona, D. Geman, N. Ikeda, École d'Eté de Probabilités de Saint Flour

XVIII, 1988. Ed.: P.L. Hennequin. VII, 330 pages. 1990.

Vol. 1428: K. Erdmann, Blocks of Tame Representation Type and Related Algebras. XV. 312 pages. 1990.

Vol. 1429: S. Homer, A. Nerode, R.A. Platek, G.E. Sacks, A. Scedrov, Logic and Computer Science. Seminar, 1988. Editor: P. Odifreddi. V, 162 pages. 1990.

Vol. 1430: W. Bruns, A. Simis (Eds.), Commutative Algebra. Proceedings. 1988. V, 160 pages. 1990.

Vol. 1431: J.G. Heywood, K. Masuda, R. Rautmann, V.A. Solonnikov (Eds.), The Navier-Stokes Equations – Theory and Numerical Methods. Proceedings, 1988. VII, 238 pages. 1990.

Vol. 1432: K. Ambos-Spies, G.H. Müller, G.E. Sacks (Eds.), Recursion Theory Week. Proceedings, 1989. VI, 393 pages. 1990.

Vol. 1433: S. Lang, W. Cherry, Topics in Nevanlinna Theory. II, 174 pages.1990.

Vol. 1434: K. Nagasaka, E. Fouvry (Eds.), Analytic Number Theory. Proceedings, 1988. VI, 218 pages. 1990.

Vol. 1435: St. Ruscheweyh, E.B. Saff, L.C. Salinas, R.S. Varga (Eds.), Computational Methods and Function Theory. Proceedings, 1989. VI, 211 pages. 1990.

Vol. 1436: S. Xambó-Descamps (Ed.), Enumerative Geometry. Proceedings, 1987. V, 303 pages. 1990.

Vol. 1437: H. Inassaridze (Ed.), K-theory and Homological Algebra. Seminar, 1987–88. V, 313 pages. 1990.

Vol. 1438: P.G. Lemarié (Ed.) Les Ondelettes en 1989. Seminar. IV, 212 pages. 1990.

Vol. 1439: E. Bujalance, J.J. Etayo, J.M. Gamboa, G. Gromadzki. Automorphism Groups of Compact Bordered Klein Surfaces: A Combinatorial Approach. XIII, 201 pages. 1990.

Vol. 1440: P. Latiolais (Ed.), Topology and Combinatorial Groups Theory. Seminar, 1985–1988. VI, 207 pages. 1990.

Vol. 1441: M. Coornaert, T. Delzant, A. Papadopoulos. Géométrie et théorie des groupes. X, 165 pages. 1990.

Vol. 1442: L. Accardi, M. von Waldenfels (Eds.), Quantum Probability and Applications V. Proceedings, 1988. VI, 413 pages. 1990.

Vol. 1443: K.H. Dovermann, R. Schultz, Equivariant Surgery Theories and Their Periodicity Properties. VI, 227 pages. 1990.

Vol. 1444: H. Korezlioglu, A.S. Ustunel (Eds.), Stochastic Analysis and Related Topics VI. Proceedings, 1988. V, 268 pages. 1990.

Vol. 1445: F. Schulz, Regularity Theory for Quasilinear Elliptic Systems and – Monge Ampère Equations in Two Dimensions. XV, 123 pages. 1990.

Vol. 1446: Methods of Nonconvex Analysis. Seminar, 1989. Editor: A. Cellina. V, 206 pages. 1990.

Vol. 1447: J.-G. Labesse, J. Schwermer (Eds), Cohomology of Arithmetic Groups and Automorphic Forms. Proceedings, 1989. V, 358 pages. 1990.

Vol. 1448: S.K. Jain, S.R. López-Permouth (Eds.), Non-Commutative Ring Theory. Proceedings, 1989. V, 166 pages. 1990.

Vol. 1449: W. Odyniec, G. Lewicki, Minimal Projections in Banach Spaces. VIII, 168 pages. 1990.

Vol. 1450: H. Fujita, T. Ikebe, S.T. Kuroda (Eds.), Functional-Analytic Methods for Partial Differential Equations. Proceedings, 1989. VII, 252 pages. 1990.

Vol. 1451: L. Alvarez-Gaumé, E. Arbarello, C. De Concini, N.J. Hitchin, Global Geometry and Mathematical Physics. Montecatini Terme 1988. Seminar. Editors: M. Francaviglia, F. Gherardelli. IX, 197 pages. 1990.

Vol. 1452: E. Hlawka, R.F. Tichy (Eds.), Number-Theoretic Analysis. Seminar, 1988–89. V, 220 pages. 1990.

Vol. 1453: Yu.G. Borisovich, Yu.E. Gliklikh (Eds.), Global Analysis – Studies and Applications IV. V, 320 pages. 1990.

Vol. 1454: F. Baldassari, S. Bosch, B. Dwork (Eds.), p-adic Analysis. Proceedings, 1989. V, 382 pages. 1990.

Vol. 1455: J.-P. Françoise, R. Roussarie (Eds.), Bifurcations of Planar Vector Fields. Proceedings, 1989. VI, 396 pages. 1990.

Vol. 1456: L.G. Kovács (Ed.), Groups – Canberra 1989. Proceedings. XII, 198 pages. 1990.

Vol. 1457: O. Axelsson, L.Yu. Kolotilina (Eds.), Preconditioned Conjugate Gradient Methods. Proceedings, 1989. V, 196 pages. 1990.

Vol. 1458: R. Schaaf, Global Solution Branches of Two Point Boundary Value Problems. XIX, 141 pages. 1990.

Vol. 1459: D. Tiba, Optimal Control of Nonsmooth Distributed Parameter Systems. VII, 159 pages. 1990.

Vol. 1460: G. Toscani, V. Boffi, S. Rionero (Eds.), Mathematical Aspects of Fluid Plasma Dynamics. Proceedings, 1988. V, 221 pages. 1991.

Vol. 1461: R. Gorenflo, S. Vessella, Abel Integral Equations. VII, 215 pages. 1991.

Vol. 1462: D. Mond, J. Montaldi (Eds.), Singularity Theory and its Applications. Warwick 1989, Part I. VIII, 405 pages. 1991.

Vol. 1463: R. Roberts, I. Stewart (Eds.), Singularity Theory and its Applications. Warwick 1989, Part II. VIII, 322 pages. 1991.

Vol. 1465: G. David, Wavelets and Singular Integrals on Curves and Surfaces. X, 107 pages. 1991.

Vol. 1466: W. Banszczyk, Additive Subgroups of Topological Vector Spaces. VII, 178 pages. 1991.

Vol. 1467: W. M. Schmidt, Diophantine Approximations and Equations. VIII, 217 pages. 1991.

Vol. 1468: J. Noguchi, T. Ohsawa (Eds.), Prospects in Complex Geometry. Proceedings, 1989. VII, 421 pages. 1991.

Vol. 1469: J. Lindenstrauss, V. D. Milman (Eds.), Geometric Aspects of Functional Analysis. Seminar 1989-90. XI, 191 pages. 1991.

Vol. 1470: E. Odell, H. Rosenthal (Eds.), Functional Analysis. Proceedings, 1987-89. VII, 199 pages. 1991.

Vol. 1471: A. A. Panchishkin, Non-Archimedean L-Functions of Siegel and Hilbert Modular Forms. V, 157 pages. 1991.

Vol. 1472: T. T. Nielsen, Bose Algebras: The Complex and Real Wave Representations. V, 132 pages. 1991.